OUTDOOR HUMAN COMFORT AND ITS ASSESSMENT

STATE OF THE ART

PREPARED BY
The Task Committee on Outdoor Human Comfort
of the Aerodynamics Committee
of the American Society of Civil Engineers

Published by the American Society of Civil Engineers

Library of Congress Cataloging-in-Publication Data

American Society of Civil Engineers. Task Committee on Outdoor Human Comfort.
  Outdoor human comfort and its assessment : state of the art / prepared by the Task
  Committee on Outdoor Human Comfort of the Aerodynamics Committee of the American
  Society of Civil Engineers.
     p. cm.
  Includes bibliographical references and index.
  ISBN 0-7844-0684-7
  1. Wind pressure. 2. Buildings--Aerodynamics--Testing. 3. Human comfort. 4. Wind
  tunnels. I. Title.

  TA654.5.A47 2003
  624--dc21   2003052401

American Society of Civil Engineers
1801 Alexander Bell Drive
Reston, Virginia, 20191-4400

www.pubs.asce.org

Contents

Task Committee Members

Edward A. Arens

Richard M. Aynsley

Leighton S. Cochran

Frank H. Durgin

Yoshihiko Hayashi

Peter A. Irwin (Chair)

Nicholas Isyumov

Shuzo Murakami

Michael J. Soligo

Theodore Stathopoulos

Hanqing Wu

1.   INTRODUCTION

Human comfort outdoors depends on a number of microclimate factors: local wind velocity, air temperature, humidity, solar radiation, precipitation, air quality and noise level. For a particular outdoor location to be comfortable these factors must fall within certain acceptable ranges. This document reviews current methods and criteria for assessing outdoor comfort.

Outdoor conditions differ from those indoors in that while the indoor microclimate can be tightly controlled for essentially 100% of the time, outside this is simply not practical. Therefore, people's expectations are somewhat different outdoors. Important factors such as local wind speeds, temperature, humidity, sun and rain are largely a function of the weather, and it is not reasonable to expect comfortable conditions 100% of the time. Thus, comfort criteria applicable to outdoors have generally been developed to include not only certain values of wind speed, temperature, solar radiation etc., but also to include statements concerning the fraction of time it is permissible for the microclimate to be outside a comfortable range, or conversely, the fraction of time that the conditions should be within a comfortable range.   Since there is considerable variability in the tolerance levels of different people (with age, health and gender being some of the variables), the criteria have also to be based on some form of population mean or on the tolerance levels of more sensitive members of the population.

An important impetus for studying the microclimate around buildings was the increasing realization in the 1960's and 1970's that a tall building can have a major impact on the wind conditions in the area immediately surrounding it. In particular a tall building that protrudes well above its surroundings tends to intercept the high velocity winds that exist higher up and deflect them down to ground level, as illustrated in Figure 1. Winds near such a building can, in extreme cases, be accelerated to several times the values that existed without it and, since wind forces vary as wind speed squared, their effect on pedestrians becomes increased by an order of magnitude. This can cause the areas at the base of the building to be uncomfortably windy and at times unsafe. There have been recorded instances of people being blown off their feet near such buildings (Melbourne, 1971and Penwarden,1973).   Figure 2 illustrates an example of an elderly woman being blown over near a tall building in the city of Calgary, Canada.

Historically, the physical action of wind has been the main focus of systematic microclimate studies around buildings. Solar radiation has been another focus but more with a view to establishing shadow patterns so as to satisfy city planning "right to light"requirements rather than for evaluating people's comfort in a systematic way. Temperature and humidity effects were recognized early on but only in more recent years has there been a concerted effort aimed at incorporating wind, temperature, humidity and solar radiation effects all into a comprehensive set of outdoor comfort criteria, Arens et al (1986), Williams et al (1992). The complexity and computational effort involved in assessing the effects of multiple microclimate factors on comfort can be substantial but available computer power has now reached a point where such assessments are much more feasible.

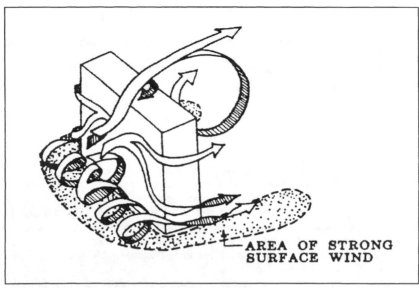

Figure 1: Wind Flow Around a Tall Building

Figure 2: Woman Blown Over

In the same period in the 1960's and 1970's that wind problems were being found around tall buildings, methods of tooting scale models of buildings in wind tunnels were being developed. The boundary layer wind tunnel, which simulates the earth's planetary boundary layer, became an essential tool for studying the wind loading on buildings and it soon became evident that it could also be used to examine the wind speeds at pedestrian level. Testing methods, along with methods for combining wind-tunnel data with meteorological statistics so as to predict the frequency of occurrence of various wind speeds at pedestrian level were developed, see Melbourne and Joubert (1971), Lawson and Penwarden (1975), Isyumov and Davenport (1975), Gandemer (1975) and Durgin (1977). Comparisons with full scale such as those of Penwarden (1973) and Isyumov and Davenport (1975) gave confidence in the wind tunnel as a predictive tool.

In parallel with the developments of wind-tunnel methods for predicting wind speeds around buildings, criteria for human comfort and safety in the presence of wind were also developed by a number of researchers in the 1970's, see Davenport (1972), Penwarden (1973), Lawson and Penwarden (1975), Gandemer (1975), Hunt et al (1976), Melbourne (1978), Murakami et al (1979). In most cases these criteria addressed purely the mechanical effects of wind action but some investigators, e.g. Davenport (1972) and Lawson and Penwarden (1975), did consider the effect of air temperature combined with wind. Lawson and Penwarden identified the important influence of clothing. Provided that people are dressed correctly for the general weather conditions of the day, the mechanical effects of wind force remain the dominant factor for comfort. In other words, people tend to automatically compensate for temperature changes using clothing to maintain thermal comfort. This finding was used to justify an approach in which thermal comfort was largely divorced from wind comfort. However, this research was focused primarily on colder climates and tended not to consider the other extreme, i.e. high temperatures, in which more wind may be desirable because of its cooling effect. Also, in extremely cold conditions, while people can keep warm by dressing appropriately, it is difficult to avoid having some exposed skin, and the avoidance of freezing of exposed skin, i.e. frost bite, needs to be considered.

In developing criteria the wind tunnel was again found to be a useful tool, in this case for observing the effects of wind on people: Hunt et al (1976), Penwarden et al (1978), Murakami et al (1981). Actual observations of people in the natural wind were also used: Melbourne (1978), Murakami et al (1979), Isyumov (1995).

The format of the wind comfort criteria put forward has varied considerably from proponent to proponent. Some have used the mean wind speed averaged over 10 minutes to an hour as the wind speed that best characterizes the impact of wind on comfort and safety. Others have employed various forms of gust speed as the best characteristic speed. Whichever definition of the characteristic speed was used, the criteria were then expressed in terms of a threshold value for this speed above which people would become uncomfortable. Also, acknowledging that comfortable conditions cannot reasonably be expected for 100% of the time outdoors, a percentage of time for which discomfort was allowed was set. For example, Lawson and Penwarden (1975) suggested that if the mean wind speeds at a particular location exceed 5 m/s for more

3

than 20% of the time then that condition is unacceptable. This was based on the experience that at projects where this criterion had been exceeded, remedial action had been initiated by the building owners in most cases. Davenport (1972) proposed criteria which varied depending on people's level of activity, being more stringent for sedentary activities, e.g. sitting, than for more strenuous activities, e.g. walking.

The question of safety was recognized as being different from comfort. The threshold speeds for safety are higher than for comfort and the frequency of occurrence has to be set at a much lower level, e.g. of order once or twice per year. For example, Melbourne (1975) proposed that exceeding a mean speed of 11.5 m/s more than about once per year was indicative of dangerous conditions. This is now considered to be rather stringent and more recent safety criteria place the threshold speed at more like 15 m/s.

During the 1980's and 1990's the use of the wind tunnel became much more wide-spread, with large numbers of projects being tested each year, and several cities instituted by-laws concerning the wind environment, e.g. Boston, Toronto, San Francisco, Auckland. Also, even where by-laws did not contain specific comfort criteria, some city planning authorities, such as in Chicago, Sacramento, London, Sydney, Melbourne and Frankfurt, were requesting wind tunnel studies of new tall buildings before granting approval for their construction. The greater use of wind tunnel studies led to an increasing number of occasions where the wind tunnel specialists had to explain comfort criteria to lay persons, which in turn showed the need to simplify the comfort criteria as much as possible. Also, it showed the need for a common set of criteria that can be used by all wind tunnel laboratories undertaking studies of proposed buildings.

It was with a view to promoting greater uniformity of methodology and criteria that this document has been prepared. It is intended to provide a review of the current methods of assessing comfort in the outdoor environment, the types of criteria that are used and methods of improving conditions at adversely affected locations. Good comfort conditions are particularly important for sensitive areas such as building entrances, outdoor restaurants, swimming pools, fountains, lounging areas, and children's play areas. They can also be important in the spectator seating and on the field of sports stadiums and in outdoor courts used for tennis or other sports. The authors of this report come mostly from a "wind" background and this will be clear from the heavy emphasis placed on the wind component. This is not inappropriate in view of the historical importance of wind to the subject and the ability that proper seasonal dress offers to moderate thermal discomfort.

However, a description is also given of recent developments aimed at including the effects of not just wind but also temperature, humidity and solar radiation. Some discussion of precipitation effects is also given. In principle, it would be possible to extend the scope of the document even further, to include the additional effects of air quality, noise and vibration. In fact, air quality is often studied using the same wind tunnel model as used to examine wind comfort (Cochran, Pielke and Kovacs, 1992). However, there is an extensive technical literature on these other factors and, as far as assessing comfort and safety is concerned, they can be largely divorced from wind and

4

thermal comfort. Therefore issues to do with air quality, noise and vibration are not considered further in this document.

2. ELEMENTS OF THE MICROCLIMATE

2.1 Wind

Wind is one of the main differences between outdoors and indoors. Even a light wind will far exceed the typical air velocities experienced inside. The wind approaching a particular site varies as large scale meteorological systems move over the earth's surface. The variations in wind speed and direction caused by these systems are gradual. It may take several days for a large cyclonic system to pass by. There are also daily variations caused by the rotation of the earth and the rising and setting of the sun. These happen over a one-day cycle. Localized disturbances such as thunderstorms may last only an hour or so. As a result, the outdoor environment is a randomly varying one, whereas the indoor environment is closely controlled.

These phenomena all cause large masses of air to move over the earth's surface. At the surface the wind is slowed by the roughness of grass, rocks, trees, hedges, buildings and numerous other sources of surface drag. Also, turbulence is created in the wind by the drag of the surface and obstacles. The end result is the formation of a turbulent planetary boundary layer in the lowest layer of the atmosphere. Within this layer the wind speed increases with height, eventually reaching an approximately constant value at the top of the boundary layer and above, Figure 3. The thickness of the planetary boundary layer can vary widely, from a few hundred to several thousand metres, ESDU (1982). The fluctuations in wind speed and direction due to the turbulence in the boundary layer occur on a very short time scale, mostly in the range of a few seconds to a few minutes.

When the wind reaches buildings, it becomes modified by them. Tall buildings tend to intercept the stronger winds high up in the boundary layer and redirect them down to ground level and this causes accelerated flow zones in front of and around the corners of the building, as illustrated in Figures 4a and 4b. If a building has an opening connecting the upwind face to the downwind face, an accelerated flow can occur in the tunnel between the two faces, Figure 4c. In gaps between buildings wind will tend to accelerate due to the Venturi effect, Figure 4d. Along with such effects buildings introduce additional turbulence into the wind. Thus, the flow around buildings is often more gusty than in the approaching winds. Figure 5 illustrates the typical variation of wind speed with time at pedestrian level, 1.5 m to 2 m above the ground surface. The wind speed in Figure 5 was measured by both a hot-wire anemometer and an Irwin Sensor (see Section 3.3 for details). People respond not only to the general or average wind flow at a location but also to the random fluctuations in speed and direction caused by turbulence. Strong winds that change suddenly can cause safety problems as people have difficulty maintaining balance.

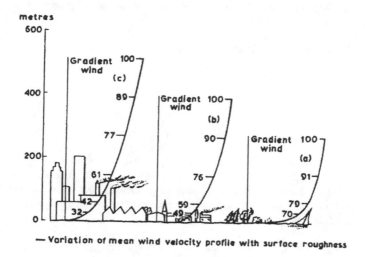

— Variation of mean wind velocity profile with surface roughness

Figure 3: Planetary Boundary Layer

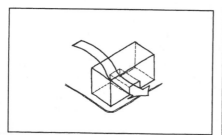

Figure 4a: Downwash

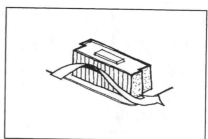

Figure 4b: Corner Stream

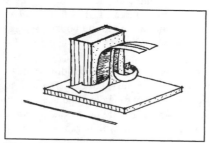

Figure 4c: Through Flow

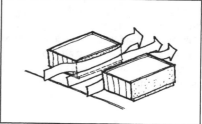

Figure 4d: Venturi Effect

6

The effects of wind on people can be divided into two categories. The first is the direct effect of wind force on the people and items such as umbrellas, dust, and loose papers. This type of effect will be loosely described as "mechanical". The second type of effect is "thermal", the more indirect influence of wind on comfort when combined with air temperature, humidity and solar radiation. The mechanical effects of various wind velocities, and the corresponding wind forces, have been described by the Beaufort Scale. The original version of this scale was developed to assist sailors and is shown in Table 1a. A modified form of this scale, applicable to wind over land, is shown in Table 1b. It is expressed in terms of mean speeds in open terrain at the 10 m height. Some additional interpretation is required to make use of the Beaufort Scale in a built-up urban area and in terms of wind velocities at the pedestrian height of about 1.5m.

As already illustrated in Figure 5, an important characteristic of the wind is that it continuously fluctuates. The upper horizontal scale in Figure 5 indicates time as experienced in the wind-tunnel test. The lower time scale illustrates typical full-scale time. There are the gradual changes in the general magnitude of the wind speed as weather systems pass over a project site, which typically occur over a day or more. However, the fluctuations shown in Figure 5 are of much shorter duration, in the range of about one second to a minute or two. The full scale fluctuations evident in this recording are due to the turbulence generated within the lowest few hundred metres of the earth's atmosphere as moving masses of air interact with the earth's surface roughness and obstacles such as buildings. They are experienced by people as gusts and lulls. Any assessment of the effects of wind on people has to acknowledge the fluctuating nature of the wind and, when formulating comfort criteria in terms of wind speeds, it is necessary to specify not only the value of the wind speed but also the averaging time used to measure that speed.

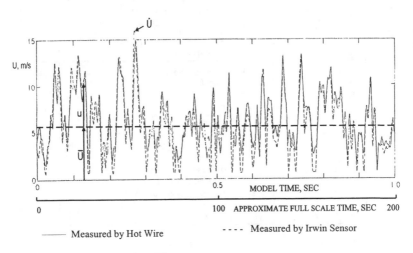

Figure 5: Typical Wind Time History

7

Table 1a: Original Beaufort Scale

The Beaufort scale of wind (nautical)							
Beaufort Number	**Name of Wind**	**Wind Speed**		**Description of Sea Surface**	**Sea Disturbance Number**	**Average Wave Height**	
		knots	kph			ft	m
0 1	calm light air	<1 1-3	<1 1-5	sea like a mirror ripples with appearance of scales are formed, without foam crests	0 0	0 0	0 0
2	light breeze	4-6	6-11	small wavelets still short but more pronounced; crests have a glassy appearance but do not break	1	0-1	0-0.3
3	gentle breeze	7-10	12-19	large wavelets; crests begin to break; foam of glassy appearance; perhaps scattered white horses	2	1-2	0.3-0.6
4	moderate breeze	11-16	20-28	small waves becoming longer; fairly frequent white horses	3	2-4	0.6-1.2
5	fresh breeze	17-21	29-38	moderate waves taking a more pronounced long form; many white horses are formed; chance of some spray	4	4-8	1.2-2.4
6	strong breeze	22-27	39-49	large waves begin to form; the white foam crests are more extensive everywhere; probably some spray	5	8-13	2.4-4
7	moderate gale (or near gale)	28-33	50-61	sea heaps up and white foam from breaking waves begins to be blown in streaks along the direction of the wind; spindrift begins to be seen	6	13-20	4-6
8	fresh gale (or gale)	34-40	62-74	moderately high waves of greater length; edges of crests break into spindrift; foam is blown in well-marked streaks along the direction of the wind	6	13-20	4-6

Table 1a cont'd: Original Beaufort Scale

The Beaufort scale of wind (nautical)							
Beaufort Number	Name of Wind	Wind Speed		Description of Sea Surface	Sea Disturbance Number	Average Wave Height	
		knots	kph			ft	m
9	strong gale	41-47	75-88	high waves; dense streaks of foam along the direction of the wind; sea begins to roll; spray affects visibility	6	13-20	4-6
10	whole gale (or storm)	48-55	89-102	very high waves with long overhanging crests; resulting foam in great patches is blown in dense white streaks along the direction of the wind; on the whole the surface of the sea takes on a white appearance; rolling of the sea becomes heavy; visibility affected	7	20-30	6-9
11	storm (or violent storm)	58-63	103-114	exceptionally high waves; small and medium sized ships might be for a long time lost to view behind the waves; sea is covered with long white patches of foam; everywhere the edges of the wave crests are blown into foam; visibility affected	8	30-45	9-14
12-17	hurricane	64 and above	117 and above	the air is filled with foam and spray; sea is completely white with driving spray; visibility very seriously affected	9	over 45	over 14

Table 1b: Beaufort Scale of Winds As Used On Land

Beaufort Number	Descriptive Term	Speed km/h	Specification for Estimating Speed
0	Calm	Less Than 2	Smoke rises vertically.
1	Light Air	2 - 5	Direction of wind shown by smoke drift but not wind vanes.
2	Light Breeze	6 - 11	Wind felt on face; leaves rustle; ordinary vane moved by wind.
3	Gentle Breeze	12 - 19	Leaves and small twigs in constant motion; wind extends light flag.
4	Moderate Breeze	20 - 29	Raises dust and loose paper; small branches are moved.
5	Fresh Breeze	30 - 39	Small trees in leaf begin to sway; crested wavelets form on inland waters.
6	Strong Breeze	40 - 50	Large branches in motion; whistling heard in telegraph wires; umbrellas used with difficulty.
7	Near Gale	51 - 61	Whole trees in motion; inconvenience felt in walking against wind.
8	Gale	60 - 74	Breaks twigs off trees; generally impedes progress.
9	Strong Gale	75 - 87	Slight structural damage occurs, e.g., to roofing shingles, TV antennae, etc..
10	Storm	88 - 102	Seldom experienced inland; trees uprooted; considerable structural damage occurs.
11	Violent Storm	103 - 116	Very rarely experienced; accompanied by widespread damage.
12	Hurricane	above 116	

The wind speed at any instant can be expressed as a mean (or average) component, $\overline{U}$, averaged over 10 minutes to one hour, plus a fluctuating component, u, as shown in Figure 5. The over-bar denotes the time average. By definition the average of u is zero. The fastest fluctuations in u that are of interest are in the range of about one to five seconds. At any instant, the total wind speed, U, can be expressed as

$$U = \overline{U} + u \tag{1}$$

One can also define a peak gust speed as being the value of the total speed U when the fluctuating component u reaches its expected maximum value, $\hat{u}$. The expected maximum value can be evaluated either as the highest over the whole hour for which $\overline{U}$ is evaluated or else as the highest that can be expected over some shorter time such as a few minutes. The former approach gives a peak gust condition that only occupies a few seconds each hour whereas the latter approach gives a more typical gust condition that would be experienced by people once every few minutes. From Equation (1) the total gust speed, $\hat{U}$, is then

$$\hat{U} = \overline{U} + \hat{u} \tag{2}$$

The peak fluctuation, $\hat{u}$, must be evaluated statistically if repeatable results are to be obtained and a number of authors have expressed it as a multiple of the root-mean-square of the fluctuations, σ_u. Then the gust speed is defined as

$$\hat{U} = \overline{U} + g\sigma_u \tag{3}$$

where g is a "peak factor". There have been a range of opinions as to what value of g gives the best characterization of the wind conditions that people feel. As indicated by Lawson (1978) values suggested for g range from 0 to 4. When $g = 0$ the effect of gusts is totally ignored. If $g = 1.5$ then the characteristic gust speed is that which is exceeded about 10% of the time. If $g = 3.5$ then the characteristic gust speed becomes the peak gust speed exceeded only about 0.1% of the time. The exact percentages corresponding to the values of g depend on the probability distribution of the wind gust speeds. 0.1% probability typically corresponds to gusts occurring a few times each hour.

Many of the effects of wind action, such as blowing dust, disturbed hair or loss of balance, tend to be primarily noticed during gusts. Therefore, some authors, e.g. Hunt et al (1976), have favoured basing criteria on the short term gusts (corresponding to $g = 3$ or higher). However, some wind effects are more dependent on the mean speed, such as the energy expended to move ahead in the face of a strong steady wind, the drying of eyes and wind cooling in low temperature conditions. The treatment of gust and mean speeds will be discussed further in the section on criteria.

2.2   Temperature and Humidity

The outdoor temperature and humidity can have a significant impact on a person's comfort. A person's sensation of comfort in cold conditions is linked to the heat balance of the human body, i.e. the balance of heat generated by metabolic processes and heat lost by conduction, convection, radiation and evaporation. Since some of the heat flows are affected by the local air speed, i.e. the convective and evaporative losses, the effects of temperature and humidity are closely linked with the wind conditions and cannot be treated in isolation from wind speed. This is why, for example, in the colder regions of North America, the wind chill equivalent temperature is used to provide a more meaningful description of how cold weather will really feel, rather than simply giving air temperature. The equivalent temperature is obtained by calculating the temperature in standard wind (set at 1.8 m/s = 4 mph) that would give the same rate of heat loss from exposed skin at 33°C as occurs in the actual wind and temperature conditions. It is estimated by first calculating a Wind Chill Index, proposed by Siple and Passel (1945), denoted here by W_{CI}.

$$W_{CI} = (10.45 + 10\overline{U}^{\frac{1}{2}} - \overline{U})(33 - T_a) \qquad \text{kcal} / (m^2 h) \tag{4}$$

where $\overline{U}$ = mean wind speed in m/s and T_a = ambient temperature in degrees C (one kcal/(m²h) = 1.162 W/m²). This is for use in the ranges $\overline{U} > 1.78$ m/s and $T_a \leq 10°C$. The wind chill equivalent temperature T_{eq} is then given by

$$T_{eq} = -0.04544 W_{CI} + 33 \qquad \text{deg. C} \tag{5}$$

For example, if the actual temperature is -10°C and the wind is blowing at 5 m/s (11.2 mph), then the wind chill equivalent temperature is calculated to be -21°C.

It is important to note that the heat loss given by Equation (4) is only for exposed skin and is primarily useful for assessing when frost-bite could occur. People tend to adjust to cold conditions by putting on more clothing which modifies the heat loss equation. Only limited areas of skin are typically left exposed in cold conditions, primarily the hands and face. Overall bodily thermal comfort will therefore be a function of not only temperature and wind speed, as in Equation (4), but also of the insulation value of clothing, metabolic rate which is a function of the level of physical activity, and exposure to solar radiation. The simultaneous consideration of all these variables is discussed further in Section 5. It should be noted that several other wind chill models are also available (Steadman, 1971; Driscoll, 1992; Bluestein and Zecher, 1999) based on other research.

It is noteworthy that the humidity of the air does not enter into Equation (4). Generally, in cold conditions, humidity is low and has little direct effect on thermal comfort, although there may be indirect effects, such as humidity changing the insulation value of clothing. In hot conditions, the human body needs to increase, not reduce, heat losses to maintain thermal comfort. This is largely achieved by reducing clothing and through sweating and the corresponding heat losses associated with the latent heat of evaporation. Since the efficiency of evaporation is decreased as the relative humidity of the air increases, the relative humidity becomes a much more important parameter in hot climates. Also, the efficiency of evaporation is increased with wind speed. Thus, while in cold climates it is often desirable to reduce wind speeds, the opposite is sometimes the case in hot climates.

The Humidex (Anderson, 1965) is an effective temperature, combining the temperature and humidity into one number to reflect the perceived temperature and to quantify human discomfort due to excessive heat and humidity (Masterton and Richardson, 1979). The Humidex, H, can be calculated using

$$H = T + h \quad \deg C \tag{6}$$

where T = dry bulb temperature in degrees C, and $h = (5/9)(e - 10)$ in which e = vapour pressure in mb or kPa $\times$ 10. The value of h or e is the moisture content of the air and can be determined from the dew point temperature, wet bulb temperature or relative humidity. Table 2 lists Humidex values based on air temperature and relative humidity. In general, almost everyone will feel uncomfortable when the Humidex ranges from 40 to 45, and many types of labour must be restricted when the Humidex is 46 and higher.

The incorporation of relative humidity effects into the overall assessment of thermal comfort is discussed in Section 5.

2.3   Solar Radiation

Solar radiation can clearly have a major impact on a person's thermal comfort. In hot climates people want to congregate in shaded areas to reduce the heat input to their bodies coming from the sun. In contrast, in cold climates the extra heat gained from solar radiation is very welcome. Therefore any assessment of thermal comfort must account for the effects of sun/shade conditions. Also, the angle of the sun, the amount of radiation absorbed by clouds, dust and particles in the atmosphere, and the sun light absorbed and reflected by buildings need to be accounted for. These topics are discussed in greater detail in Section 5 and Appendix A.

2.4   Precipitation

It is difficult to avoid rain, snow, hail or sleet if you are outside. Therefore consideration of the effect of precipitation on comfort tends to be of lesser interest compared with other microclimate factors except where a canopy or roof is being specifically designed to provide overhead shelter. In this case, it is of interest to identify how far under the canopy the precipitation will infiltrate and how often this will occur. Also, in situations where people travel in rainy conditions to a baseball game for example, where they may be under a canopy, the dampness of their clothes could have an effect on their thermal comfort.

Table 2: Humidex

| Table - Humidex from temperature and relative humidity readings |||||||||||||||||| |
Temp (oC)	100	95	90	85	80	75	70	65	60	55	50	45	40	35	30	25	20
43	Humidex (oC) — Degree of Comfort												56	54	51	49	47
42	20 - 29 Comfortable											56	54	52	50	48	46
41	30 - 39 Varying degrees of comfort										56	54	52	50	48	46	44
40	40 - 45 Almost everyone uncomfortable / 46 and over Many types of labour must be restricted									57	54	52	51	49	47	44	43
39									56	54	53	51	49	47	45	43	41
38							57	56	54	52	51	49	47	46	43	42	40
37					58	57	55	53	51	50	49	47	45	43	42	·40	
36			58	57	56	54	53	51	50	48	47	45	43	42	40	38	
35		58	57	56	54	52	51	49	48	47	45	43	42	41	38	37	
34	58	57	55	53	52	51	49	48	47	45	43	42	41	39	37	36	
33	55	54	52	51	50	48	47	46	44	43	42	40	38	37	36	34	
32	52	51	50	49	47	46	45	43	42	41	39	38	37	36	34	33	
31	50	49	48	46	45	44	43	41	40	39	38	36	35	34	33	31	
30	48	47	46	44	43	42	41	40	38	37	36	35	34	33	31	31	
29	46	45	44	43	42	41	39	38	37	36	34	33	32	31	30		
28	43	42	41	41	39	38	37	36	35	34	33	32	31	29	28		
27	41	40	39	38	37	36	35	34	33	32	31	30	29	28	28		
26	39	38	37	36	35	34	33	32	31	31	29	28	28	27			
25	37	36	35	34	33	33	32	31	30	29	28	27	27	26			
24	35	34	33	33	32	31	30	29	28	28	27	26	26	25			
23	33	32	32	31	30	29	28	27	27	26	25	24	23				
22	31	29	29	28	28	27	26	26	24	24	23	23					
21	29	29	28	27	27	26	26	24	24	23	23	22					

Legend (Degree of Comfort):
- 20 - 29: Comfortable
- 30 - 39: Varying degrees of comfort
- 40 - 45: Almost everyone uncomfortable
- 46 and over: Many types of labour must be restricted

One other factor relevant to comfort assessment is that when it is raining heavily people are less likely to be outside in plazas or outdoor leisure areas. Thus, their wind and thermal comfort will usually be of less importance in rainy conditions.

3. ASSESSING AND CONTROLLING WIND COMFORT

3.1 Wind Statistics

As indicated in the introduction, comfort and safety criteria for the effects of wind on pedestrians must be stated in terms of the fraction of time that conditions are either comfortable or not comfortable, or that they are unsafe. Therefore it is imperative to know the local wind statistics, in particular the probability distribution of the mean wind speed and direction at an appropriate reference location not too far from the site. Usually wind records are available from a nearby airport. Either the complete hourly history of these records can be used directly with the wind tunnel data in the so-called time history approach, or they can be used to develop a statistical model of wind for the area. The increases in computing power that continue to take place have increased use of the former approach in recent years but historically the majority of studies have employed the latter approach. A commonly used mathematical expression that models the statistics fairly well in most cases is the Weibull expression, which can be stated as

$$P(\overline{U},\theta) = A(\theta)\exp\left(-\left(\frac{\overline{U}}{c(\theta)}\right)^{k(\theta)}\right) \tag{7}$$

where $P(\overline{U},\theta)$ is the probability that the mean wind speed exceeds the value $\overline{U}$ for the wind directions within an azimuthal sector of selected size $\Delta\theta$ centred on the direction θ. The factor $A(\theta)$ is the fraction of the time that the wind blows from the selected sector, and $c(\theta)$ and $k(\theta)$ are constants required for each of the direction sectors. Typically $\Delta\theta$ is either 22.5 degrees (i.e. one sixteenth of the compass range), or 10 degrees (one thirty-sixth of the compass range).

Figure 6 shows an example of a typical fit of the Weibull curve to the wind data for a selected wind sector. Note that the "single fit" curve in Figure 6 has been fitted with emphasis on the lower speeds. A second Weibull curve may be required to better fit the high-speed portion of the wind data as illustrated by the "double fit" curve in Figure 6. Figure 7 shows an example of a plot of the $A(\theta)$ coefficient which illustrates the fraction of the time the wind blows from various directions. The directionality of the wind can be very important. A wind-tunnel model may exhibit very high wind speeds at a particular location for a certain wind direction. If the wind direction is one that occurs frequently then this location probably will be one with unsafe or uncomfortable winds, representing a potential problem. If it is a direction which occurs infrequently then there is probably no problem.

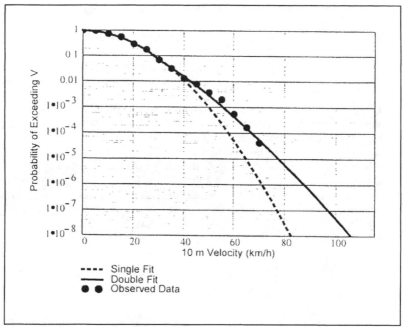

Figure 6: Wind Statistics Weibull Fit

The standard height for meteorological measurements is 10 m above ground and ideally the anemometer used to make the measurements is in open terrain, unaffected by nearby obstacles such as buildings or trees. Although most anemometers at airports come close to being at an ideal site, it is rare that the surrounding terrain corresponds perfectly to genuine open terrain conditions or that the height is exactly 10 m. Therefore it is typically necessary to make some corrections to the raw data or to the fitted mathematical model to take account of the anemometer location and height. For example, airports often have suburbs nearby and the open fetch on the airfield itself is frequently insufficient for the effects of the rougher suburban terrain to have disappeared completely by the time the wind reaches the anemometer. Methods such as those described in ESDU(1982, 1983, 1985, 1993) and Cook (1985) can be used to evaluate the effects of such terrain changes on the wind velocity profile. The end result that is needed is the probability distribution of the winds at gradient height since it is gradient height information that is eventually combined with wind-tunnel data. Figure 3 indicates typical values of gradient height assumed in wind tunnel studies. A Weibull model fitted to ideal open terrain 10 m height data can readily be converted to gradient height by simply factoring up the $c(\theta)$ constants by the ratio of gradient wind velocity to 10 m velocity. For example, the mean velocity profile within the boundary layer in open terrain can be reasonably represented by the power law

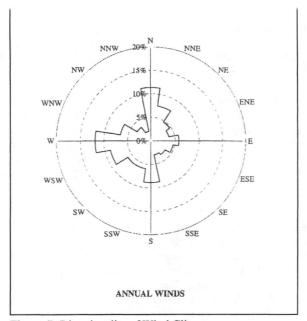

Figure 7: Directionality of Wind Climate

$$\frac{\overline{U}_g}{\overline{U}_{10}} = \left(\frac{z_g}{10}\right)^{1/7} \tag{8}$$

where $\overline{U}_g$ and z_g are the gradient wind velocity and gradient height respectively and $\overline{U}_{10}$ is the 10 m height wind speed. As an example, this expression gives ratios of $\left(\overline{U}_g / \overline{U}_{10}\right)$ of 1.63 and 1.80, for gradient heights of 300 m and 600 m respectively. In suburban terrain the exponent for the mean velocity profile will tend to be in the range 1/5 to 1/4.

3.2   Wind Tunnel and Water Flume Methods

3.2.1   General

The flows around buildings in complex urban settings are still extremely difficult to predict with any accuracy by computational methods, even with currently available computing power. However, the testing of scale models in a wind tunnel has, since the 1960s, been shown to be a very effective method of prediction. The wind-tunnel model typically includes all buildings in the surrounding landscape. Thus, their effect is automatically included. Both existing conditions and those with the new project in place can be readily measured, thus allowing the impact of the new project to be identified. Furthermore, where undesirable wind conditions are found, the effects of changes to the building itself, or to landscaping, can also be studied.

In principle, water flumes can achieve the same results provided the dimensions of the water flume and water depth are sufficient compared with the model to generate an accurate simulation of the planetary boundary layer, but the wind tunnel has become by far the preferred tool in practice. It is important that the wind tunnel (or water flume) be capable of simulating the mean-velocity profile and turbulence of the natural wind, or unrepresentative results may be obtained.

3.2.2 Wind Simulation

A typical set up of a wind-tunnel model in a boundary-layer wind tunnel is illustrated in Figure 8. The building itself and the model of its surroundings are mounted on the wind-tunnel turntable which can be rotated to allow various wind directions to be simulated. Typical model scales for large buildings are in the range of 1:200 to 1:500. Larger scales have been used for smaller buildings. The model of surroundings enables the complex flows created by other buildings near the study building to be automatically included in the tests. However, it is also essential to create a proper simulation of the natural wind approaching the modelled area. The requirements for modelling the natural wind in a wind tunnel are described in the *ASCE Manual of Practice No.67 on Wind Tunnel Studies of Buildings and Other Structures,* (ASCE 1999). The mean-velocity profile, the turbulence-intensity profile and the integral length scale of the turbulence in the boundary layer on the wind-tunnel floor must all be representatively matched to full-scale values. This is typically achieved by using a long working section with a specially roughened floor upwind of the model, as can be seen in Figure 8. Often this is supplemented by having tapered spires near the upwind end of the working section that are shaped to slow the wind more near the floor and to introduce large scale turbulence. Full details of the various methods used can be found in the ASCE Manual of Practice referenced above.

The reference wind speed for the tests is usually taken to be the gradient wind speed $\overline{U}_g$ above the boundary layer. All speeds measured around the model are divided by this reference speed which converts them to speed ratios. Since these speed ratios are essentially independent of the actual reference speed for the tests, the choice of reference speed is, to some, extent arbitrary, and can be optimized for the wind-tunnel instrumentation used. However, very low reference speeds should be avoided in order to maintain measurement resolution and to keep the wind-tunnel Reynolds number high enough to avoid the onset of viscosity effects in the air flow. In typical wind tunnel tests, the air flow speed above the boundary layer is in the range 10 to 30 m/s.

Figure 8: Boundary Layer Wind Tunnel Test

3.3   Measurement Techniques

3.3.1   Basic Requirements

In a typical boundary-layer wind-tunnel test, measurements of mean wind speeds at various locations near the ground, as well as gust speeds and rms speeds, are normally the minimum requirements for the evaluation of wind conditions. Probability distributions of wind speeds, local wind directions and direction variations may also be used in some cases. The measurements must sample the wind for a sufficient time to obtain statistically stable values of quantities such as mean, rms and peak values. Clearly, the more locations that are measured the better. The number of measurement locations will depend on the extent of the area to be covered and on the type of instrumentation used. When hot-wire measurements are taken, the number of locations may be in the range 20 to 40, for example. When Irwin sensors (see Section 3.3.2) are used, typically more locations are feasible and it is not uncommon for the number to be in the range 50 to 100. Some studies have used several hundred (Cochran and Howell, 1990; Williams and Wardlaw, 1992). Figure 9 illustrates wind speed sensors distributed around a wind tunnel model.

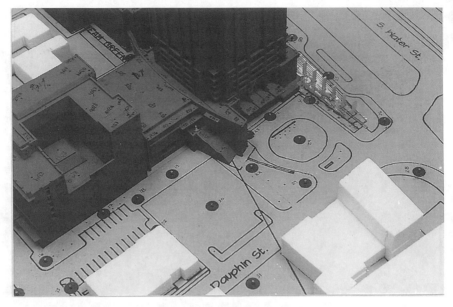

Figure 9: Irwin Sensors Installed Around A Wind Tunnel Model

The number of test locations generally possible is likely to increase in the future as instrumentation and data acquisition systems continue to improve. The instrumentation should create minimum disturbance to the wind flow. The choice of experimental technique must be guided by the requirements for accuracy, repeatability, stability, resolution and cost.

The fact that only wind conditions near the ground have direct effects on pedestrians necessitates a low measurement height. Commonly used heights are in the range 1.5 to 2 m above the ground at full scale, corresponding to 4 to 5 mm in a simulated boundary-layer flow when using a typical 1/400 geometrical scale. At such a low level, the wind flow tends have a low mean speed, a high turbulence intensity and a high shear stress. As indicated earlier, gust durations in the 1 to 5 second range are of interest for pedestrian comfort and safety. The frequency response required for instrumentation can be estimated from the geometric and speed scales of the wind tunnel test. Typically this results in the requirement for a flat response up to frequencies in the range of about 50 to 100 Hz. Measurements are usually taken at 10 or 22.5 degree increments of wind azimuth and may be repeated with and without the new project present. Further repetitions of the measurements may be made to examine the effects of landscaping, wind screening devices and/or modifications to the building.

The following sections discuss some typical wind-tunnel measurement techniques for pedestrian-wind assessment. Additional details can be found in Ettouney and Fricke (1975), Irwin (1982), Beranek (1984), and Wu and Stathopoulos (1993). Before describing these techniques it should be pointed out that there are two classes of measurement that can be defined in pedestrian wind studies: point methods and area methods. Ideally one would be able to determine the wind speed at pedestrian level as a continuous function of position on the ground. Methods that try to achieve this type of result are termed area methods. However, bearing in mind that the speed is a rapidly varying function of time and that we want to know the peak gusts as well as the mean value, the area approach turns out to be difficult to execute in practice without sacrificing measurement precision. What has been found to be more achievable is the measurement of wind speed, including both mean and gust values, at specific points on the wind-tunnel model. These methods are termed "point" methods. In practice, point methods tend to be used much more widely and routinely than area methods. Ultimately there may well be a convergence of these two approaches as improved instrumentation allows a larger and larger number of points to be studied in the "point" methods so that they approach the point density required to achieve an "area" type result. The following sections summarize the main point and area methods. Further details and references on these methods can be found in Appendix B.

3.3.2   Point Methods

Hot-wire and Hot-film Anemometry: As one of the most reliable and widely-used methods for flow measurements, thermal anemometry, using hot-wire or hot-film sensors, has generated a large volume of literature, e.g., Schubauer and Klebanoff (1946); Sandborn (1972); Fingerson and Freymuth (1983). Figure 10 shows an example of a hot-film probe and a hot-wire probe looks very similar. The hot wire is small in size and high in frequency response (up to 10,000 Hz) within a wide velocity range. However, it does have some drawbacks for pedestrian-wind measurements, because of its fragility, sensitivity to temperature change in the environment, and changes in calibration caused by accretion of dust particles. From this viewpoint, the hot film is superior since it is more robust, although it has a larger size and lower frequency response (in the range of 100 Hz to 200 Hz, depending on the probe configuration) - which is, however, quite adequate for the assessment of pedestrian-level winds.

During a test, a single-wire (or film) sensor is usually positioned vertically in the wind tunnel. It measures the horizontal wind component and provides an average speed over the sensor length. If the turbulence intensity to be measured is greater than 30%, thermal anemometers may provide high mean and low rms velocities, mainly due to the non-linear calibration relations. Hot-wire and hot-film anemometers are in routine use at a number of laboratories.

Figure 10: Photo of Hot-Film Probe

Irwin Sensor: Irwin (1981) proposed a simple pressure sensor, illustrated in Figure 11, which has a number of advantages in pedestrian-level wind studies. The sensor consists of circular tube, usually made from hypodermic tubing, protruding vertically from a flat base. At the base of the tube is an annular hole of slightly larger diameter than the outside diameter of the tube, and extending at least 1.5 tube diameters below the surface. A pressure tap is installed at the bottom of the hole. The pressure difference between the top of the tube and the tap at the bottom of the annular hole is measured and is converted to a wind speed at the level of the top of the tube using a pre-determined calibration function. No alignment of the sensor with the wind direction is required since, with the above geometry of the sensor, its output is independent of wind direction. Essentially it is used to measure local wind speed, not direction, in the same way that hot-film or hot-wire instrumentation is typically used in pedestrian level wind studies. It is the impact of the local wind speed that counts most for pedestrians and the local direction is of much less importance. Irwin (1981) demonstrated, by comparison with hot-wire measurements, that the sensors were able to measure not only the mean speed but also speed fluctuations. The frequency response was found to be high enough to cover the range of interest for pedestrian wind studies.

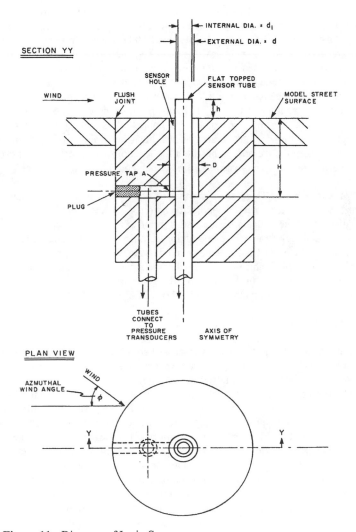

Figure 11: Diagram of Irwin Sensor

Unlike the thermal anemometer, the pressure sensor is sturdy, and relatively easy and inexpensive to manufacture. However, manufacturing must ensure high geometric accuracy. A large number of sensors can be installed on a model site, allowing this method to more closely approach an "area" method than the hot-film/hot-wire approach. Limiting factors on the number that can be installed are the available number of pressure measurement channels (which can be several hundred), and the need to avoid excessive flow interference between nearby sensors. The pressure measuring system must be sensitive and care must be taken to eliminate zero drift because the pressure signals from the sensors become very low in low velocity regions. Irwin (1981) and Wu and Stathopoulos (1994) examined interference effects and found they become insignificant for longitudinal and lateral sensor separations greater than about twelve and five times the sensor tube diameter, respectively. The sensor can in principle be used to determine wind speeds somewhat above the top of its tube provided adjustments are made to the calibration formula. However, Wu and Stathopoulos (1994) and Durgin (1992) found that in regions of strong down flow, such as near the windward face of a tall building, it is advisable to keep the calibration height at the level of the sensor top. The pressure technique using the Irwin sensor is in routine use at a number of laboratories.

Other Point Methods There are a number of other point methods that have been experimented with such as laser-doppler techniques, particle-velocimetric methods, small-scale wind-force indicators and various forms of skin friction device. However, none have so far demonstrated the combination of utility, simplicity and cost of the hot-wire, hot-film or the Irwin sensor methods and thus are not in widespread routine use in pedestrian-wind studies. This is not to say that the situation could not change in future as alternative methods are further developed. The interested reader can find further discussion and references on other point methods in Appendix B.

3.3.3 Area Methods

Flow Visualization Using Smoke As discussed above, point methods only provided quantitative data at specific points. To a large extent, this shortcoming can be mitigated by making the flow visible by the introduction of non-toxic smoke into the flow around the model buildings in the wind tunnel. This gives the observer a good feel for the extent of accelerated flow zones and is a very common technique, typically combined with point methods. Various types of smoke generator are available commercially. Many of the important wind flows influencing the comfort of pedestrians at street level can be identified by observing the smoke flow patterns. This type of visual presentation is particularly helpful for lay persons, not familiar with aerodynamics. Down-flows, accelerated flows, rolled up vortices in front of a building and the horseshoe vortices curling around the sides can all be made visible in their full, three-dimensional complexity. Figure 12 gives an example of smoke flow visualization in low turbulence. The visualization becomes more irregular in the highly turbulent flows found in urban settings.

Figure 12: Wind Flow in Front of a Tall Building (wind blowing from left to right)
©BRE. Photograph reproduced here by kind permission of Building Research Establishment.

Erosion and Surface Flow Visualization Techniques The erosion test identifies the windy zones by scour patterns in particles distributed over the model. Appendix B provides some details of these techniques. Typically the wind speed is increased in steps and photographic or digital image processing techniques are used to identify changes in the scour patterns observed at different speeds (Borges and Saraiva, 1979, and Livesey et al, 1990). The greatest advantages of this technique are its continuous coverage of large areas and its ability to provide visualization of the flow field. However, there are difficulties in relating the scour patterns to specific quantities such as mean or gust speeds. Also, carrying out tests for a number of wind directions and cleaning the work area following each test make this method labor-intensive. In practice, this has prevented its use on a routine basis. However, it can be beneficial for early exploratory studies of large areas of development.

Surface flow-visualization techniques follow principles similar to those of the erosion technique but the visualization materials are fluid mixtures, like paraffin oil with kaolin powder, pigment paint and other materials. This type of visual presentation again helps architects and city planners more readily understand the wind flow patterns influencing the comfort or safety of a particular area. Other visualization techniques, like smoke streaklines emitted from holes in the model, particle injection, tufts and directional vanes (point methods), can also be effective.

3.4 Computational Methods

Knowledge-Based Computer Systems There are many building projects where some knowledge of the building's influence on the wind flows around it and of methods of improving them would be useful in the early stages of design. Precise knowledge may not be important at this stage. Also there are smaller projects where detailed wind tunnel studies would not be warranted for cost and timing reasons. In these cases wind consultants may be approached to give the benefit of their experience as experts, not to undertake a full study involving wind-tunnel tests. Some research has been undertaken to examine the possibility of developing knowledge-based expert computer systems that could mimic the expert. At present the development of such systems is still evolving and they are not in routine use. A description of research on knowledge-based systems is given in Appendix B.

Computational Wind Engineering (CWE) Computational Wind Engineering (CWE) is a new branch of Wind Engineering in which Computational Fluid Dynamics (CFD) methods are used to compute the detailed flow patterns around buildings. Thus, the computer essentially replaces the wind tunnel with this approach, at least in principle. However, these methods involve very large amounts of computation even for relatively simple problems and their accuracy is often difficult to assess when applied to a new problem where prior experimental verification has not been done. A key difficulty is the problem of modelling of wind turbulence with any degree of generality, and wind turbulence is extremely important in the context of exterior building aerodynamics. This has persistently defied solution despite extensive research efforts over many decades. Thus, while impressive graphic displays of wind flow patterns can be generated by CFD methods, until further advances in turbulence modelling are made and verification obtained, their accuracy is in doubt. For these reasons the application of CFD to exterior wind flow problems of buildings has to date been primarily for making preliminary evaluations of the wind flows around a project, and when used in this mode they can be very useful. For final definitive results wind-tunnel testing is still required. In the future, as computing power continues to increase, the role of CWE can be expected to become more prominent. Increased computing power allows more realistic, and generally applicable ways, of simulating turbulence to be used, e.g. the large eddy simulation method in which only the small eddies of turbulence need be modelled approximately and the large eddies are computed in detail. Appendix B gives further details of research on CWE methods.

4.   WIND CRITERIA AND CONTROL MEASURES

4.1   Criteria for Comfort in Terms of Wind Force

As discussed in the introduction, a variety of criteria for comfort have been proposed by a different researchers. Appendix C summarises a number of these. It is not the intent of this document to exhaustively review all these criteria, some of which were developed many years ago. Rather, the intent is to reflect more recent trends and to respond to the need to keep the criteria simple and readily understood by lay persons.

Section 2 described how the wind speed varies with time and explained the need to define a characteristic wind speed. Gust speeds are important in some circumstances, particularly where winds are very strong and people's balance is involved, but mean speeds might be just as important an indicator of comfort in areas such as outdoor lounging areas. Therefore, it is desirable for the criteria to capture the effects of both mean speeds and gust speeds. A way of achieving this (Durgin, 1995, Lawson 1978) is to express the criteria in terms of mean speeds, but then to compare not just the mean speed $\overline{U}$ but also a "Gust Equivalent Mean" speed U_{GEM} with these criteria. "The Gust Equivalent Mean" speed U_{GEM} is computed as

$$U_{GEM} = \frac{\hat{U}}{G} \tag{9}$$

where $\hat{U}$ is the peak gust speed discussed in Section 2, and G is a representative fixed gust factor. It is suggested that $\hat{U}$ be the peak 3-second long gust exceeded about once every 5 to 10 minutes which is typically similar to assuming a peak factor g in Equation 3 of about 3.5.

Thus, with this approach, two characteristic speeds, $\overline{U}$ and U_{GEM}, are computed and both are compared with the same criteria. If the criteria are exceeded by one or both then this indicates an overall assessment of "uncomfortable". The value of G is typically taken as about 1.85.

Since it is unreasonable to expect comfortable conditions for 100% of the time, most criteria are specified in terms of wind speeds that are exceeded for some percentage of the time and the threshold wind speeds for discomfort are varied depending on the human activity anticipated at the location in question. For example, lower wind speeds are needed in areas where people might sit, such as at an outdoor restaurant, compared with areas where people are walking, en-route to some other location. Intermediate criteria for wind speed would apply where people are expected to stand for a while, such as at a bus stop or building entrance. Some criteria have gone further and subdivided walking into strolling and walking purposefully, also called business walking. In fact a variety of categories have been proposed as can be seen in Appendix C. An important part of developing criteria is that they reflect the available subjective data coming from studies on people's responses such as those by Murakami and Deguchi (1981) and Isyumov (1995) and others referred to in the introduction.

Concerning the percentage of time that conditions are permitted to be uncomfortable, this has ranged from 1% to 20%. To a large extent, the choice of percentage is arbitrary in that one can set the threshold wind speeds for "discomfort" at higher values but reduce the percentage of the time that they are permitted to be exceeded. By adjusting the threshold wind speeds to compensate for different percentage levels one can obtain similar end results. However, experience in public hearings indicates that lay persons have difficulty understanding that there is a problem if the comfort criteria are only exceeded for a small percentage of the time. For example, if 1% is the chosen percentage for the criteria, and a particular location fails the criteria, the developer would often respond: "But this means my project gives comfortable conditions for 99% of the time. So what's the problem?". This type of confusion is much less likely to occur if the threshold wind speeds are set a little lower and the percentage of time that the threshold wind speeds are permitted to be exceeded is set at 20% say, or, at a minimum 5%. Also, in the interests of avoiding confusion amongst lay persons, it appears best to avoid employing different percentages for different activities, as has been proposed by some authors.

Table 3 gives an example of a simple set of criteria. It lays out the suggested criteria for three levels of activity, sitting, standing and walking, and is based on 20% probability of exceedance. Criteria similar to these have been used on many projects in North America.

Table 3:   Example of Simple Criteria, Based on 20% Probability of Exceedance

Activity	Comfortable Ranges for $\overline{U}$ and U_{GEM} , m/s
Uncomfortable for any activity	> 5.4
Walking	0 - 5.4
Standing	0 - 3.9
Sitting	0 - 2.6

Note:- 1 m/s = 2.24 mph

If the wind speed is within the ranges in Table 3 for 80% of the time or more then conditions are said to be "comfortable" for the given activity. If it fails to stay within the 0 to 5.4 m/s range for at least 80% of the time then the conditions are said to be "uncomfortable for any activity". It should be noted that 80% is typically what would be used but, depending on the use of the area, other percentages may be appropriate in some cases.   For example, at a rarely used location, requiring 80% might be unnecessarily stringent and 70% or 60% may be agreed upon as more reasonable.

Table 4 gives an example of a slightly more detailed set of criteria with four levels of activity rather than three. These criteria were used, for example, on the Canary Warf development in London, England, as well as other projects, and are based on 5% probability of exceedance. Approximate speed ranges corresponding to 20% probability have also been given in the final column and it can be seen that the lowest three comfort levels C2, C3, and C4 are equivalent to the three comfort categories of Table 3, Walking, Standing and Sitting respectively. The difference between Table 4 and Table 3 is therefore the introduction of one additional category, business walking which is more permissive than the Walking category of Table 3.

Table 4: Example of More Detailed Criteria, Based on 5% or 20% Probability of Exceedance

Comfort Level Guideline	Activity	Comfort Ranges for $\overline{U}$ and U_{GEM} 5% probability	Description of Wind Effects	Approximate Corresponding Ranges for $\overline{U}$ and U_{GEM} at 20% probability
C1+	Exceeds Comfort Criteria	> 10 m/s	• Umbrellas used with difficulty • Hair blown straight • Difficult to walk straight • Wind noise on ears unpleasant	>6.8 m/s
C1	Walking Purposefully or Business Walking	0 - 10 m/s	• Force of wind felt on body • Trees in leaf begin to move • Limit of agreeable wind on land	0 - 6.8 m/s
C2	Strolling or Window Shopping	0 - 8 m/s	• Moderate, raises dust, loose paper • Hair disarranged • Small branches move	0 - 5.4 m/s
C3	Standing or sitting - short exposure	0 - 6 m/s	• Hair is disturbed, clothing flaps • Light leaves and twigs in motion • Wind extends lightweight flag	0 - 3.9 m/s
C4	Standing or Sitting - long exposure	0 - 4 m/s	• Light wind felt on face • Leaves rustle	0 - 2.6 m/s

Note:- 1 m/s = 2.24 mph.

It must be recognized that there is uncertainty associated with comfort criteria. This is evidenced by the differences in the threshold speeds proposed by various researchers, see Appendix C. Different people subjected to wind will respond in different ways and this may well be influenced by physical robustness, age, gender, health, size etc. and well as the tendency to adapt to the general windiness of the region. A rough estimate of the uncertainty in the speed criteria for any given level of probability is that it is probably in the range of 10% to 15%. In view of this, there are limits to the level of sophistication that the criteria can realistically be taken to. The examples of comfort criteria in Tables 3 and 4 are probably about as sophisticated as can be justified based on current knowledge and are sufficiently simple for most lay people to readily understand. It is important to take every opportunity to check the applicability of the criteria to particular regions by making use of information on locations identified by locals as windy. These can serve as useful reference points. Regional differences may require adjustments to the criteria in some cases.

4.2   Criteria for Safety

As discussed in the introduction, the winds around some tall buildings have been sufficient to blow people over. This is clearly unsafe. The gust speeds that are enough to cause people to be blown over have been estimated by various authors and are in the range of 20 to 30 m/s (45 to 67 mph), depending on a number of factors including the size, reaction time, clothing and health of the person involved. A representative value is in the middle of this range, i.e. 25 m/s or 56 mph. If wind conditions with this gust speed are exceeded more than two or three times per year then the chance of someone being injured becomes unacceptably high. Two or three times per year corresponds to events that occur for about 0.1% of the time. Thus, to satisfy the requirement for safety it is suggested that wind conditions with peak 3-second gusts exceeding 25 m/s should not occur for more than 0.1% of the time. Note that some practitioners prefer to use a frequency of occurrence of once per year as the design safety condition which, if 25 m/s is retained as the threshold speed, is slightly more stringent in most cases.

4.3   Methods of Controlling the Wind Environment

4.3.1   Building Massing and Orientation

A windy environment around the base of a building, particularly near a main entrance or plaza area will detract from the appeal of the site and perhaps discourage clients and shoppers from visiting the area. Many examples exist of outdoor restaurants and cafes that have failed at the base of tall buildings as a result of a windy environment (Cochran, 1979). Similarly, an outdoor pedestrian space, such as a recreational pool area of a residential condominium, should be protected from strong winds. Thus, there is a direct financial motivation for ameliorating the wind environment if it is going to effect the appeal of a building to the users and customers of that building. Also, a growing concern is the risk of litigation by individual citizens or groups, should a particular project cause hazardous conditions in public areas.

It is well known that the design of a building will influence the quality of the ambient wind environment at its base. A shear curtainwall to ground level with a rectilinear floor plan is often a design which can aggravate street-level winds by allowing the high-elevation, faster winds to flow down the face of the structure (circular shapes typically do not cause flows of this type). The mechanism is called downwash (see Figure 4a) and the resulting flow is then accelerated around the ground level corners (see Figure 4b). The flow may be interrupted by a large canopy over the building entrances. This is a common architectural feature used as a mitigation measure on many major projects in recent years.

Another massing issue which may be a cause of strong ground-level winds is an arcade or thoroughfare opening from one side of the building to the other. This effectively connects a positive-pressure region on the windward side with a negative-pressure region on the lee side. A strong flow through the opening often results, as illustrated in Figure 4c. A similar phenomenon occurs with a high-rise building raised up on columns. The uninviting nature of these open areas is a contributing reason behind the rarity of this type of architectural massing in modern high-rise buildings. One exception is the use of this configuration in calm, tropical climates, where the extra breeze creates a desirable feature. For example, several structures in Singapore use this approach to provide shaded, cooler areas at the ground-level entrances.

Horizontally accelerated flow between two buildings (see Figure 4d) is another cause of a windy ground-level pedestrian environment, which may also be locally aggravated by ground topography. By inspection of the available wind data, the designer may find a dominant wind direction that can be used to orient the building on the site so as to minimize these accelerated flows in highly trafficked pedestrian areas. In general, the relative position of an individual building with respect to other surrounding buildings has important consequences for the winds at ground level (see for example: Isyumov, 1995; Irwin, Hunter and Williams, 2001). Therefore, some consideration needs to be given to the potential effects of other buildings that might be built near the study building in future, especially in rapidly developing cities.

The way in which a building's vertical line is broken up may also have an impact. For example, if the floor plans have a decreasing area with height the flow down the windward face may be greatly diminished. To some extent, the presence of many balconies can also mitigate the impact on pedestrian-level winds. However, designs with many elevated balconies and deck areas near a building corner may cause a windy environment at those locations. Mid-building balconies are usually calmer.

In summary, there are two principal types of flow that adversely affect the pedestrian environment: (i) downwash flows that bring faster moving wind from higher to lower elevations (usually best diminished by a podium, or large canopy, or a stepped back design), and (ii) horizontally accelerated flows (often ameliorated by porous screens or plantings). These approaches are discussed below.

31

It is worth commenting that while tall buildings can cause undesirable increases in wind at pedestrian level on windy days, they can also help ventilate street canyons in light wind conditions, which helps to reduce air pollution from traffic and other sources. In heavy traffic areas of large cities it is desirable to maintain some minimum level of ventilation, see Isyumov (1993), Cochran and Howell (1990) and Heidorn, Murphy and Davies (1989) for further discussion of this issue.

4.3.2 Architectural and Landscaping Features

Once the flow mechanism at a problem location has been established for the critical wind directions the remedial solutions may be explored for effectiveness in a boundary-layer wind tunnel. As noted earlier, downwash off a tower may be deflected away from pedestrian areas by large canopies or podium blocks. The latter have become popular lately, with the main tower being set back from the edge of a larger full-site podium of several floors in height. The downwash then effectively impacts the podium roof rather than the public areas at the base of the tower. Provided that this podium roof area is not intended for recreational use (e.g. swimming pool, tennis court, putting green etc), this massing method is typically quite successful.

Horizontally accelerated flows that create a windy environment are best dealt with by using porous screens or substantial landscaping. Large hedges, bushes or other porous media serve to break up and retard the flow and dissipate the kinetic energy available to cause a windy environment. A solidity ratio (i.e. proportion of solid area to total area) of about 50-70% has been shown to be most effective in reducing the flow's momentum over substantial areas downwind. Solid barriers provide smaller regions of protection and produce greater accelerations at their edges.

These physical changes to the pedestrian areas are most easily evaluated by a model study in a boundary-layer wind tunnel. A comparative study showing the impact on the site with and without the ameliorative additions is a useful method of defining their effectiveness. When these studies are done as part of the design process the architect can be more assured of the successful operation of the open spaces and entrances around a new project.

5.   ASSESSING THERMAL COMFORT

5.1   Thermal Models of the Human Body

To evaluate people's thermal comfort is more complex than evaluating the effects of the mechanical forces of wind. Feelings of comfort are related to changes in body temperature due to excessive heat loss or excessive heat gain. It is the large number of variables that affect the rate of heat loss or gain that makes the problem complex. The variables that affect heat gain or loss rates are air temperature, relative humidity, solar radiation, wind speed, clothing, the activity the person is engaged in, exposure time and physiological regulatory mechanisms (such as sweating) that can to a large degree compensate for changes in the other variables so as to maintain comfort. In addition, the physiological regulatory mechanisms and heat transfer coefficients for the body are subject to variation from person to person due to the inherent variability of physical characteristics amongst the population. Age, fitness and aerobic capacity, lean and fat body mass and gender all have an influence.

To assess thermal comfort in a systematic way requires that the heat generation and heat loss processes be examined in detail and to this end mathematical models of these processes have been developed. Two well known examples of these models are the Fanger model (1972) and the Pierce Two-Node model (Gagge, Fobelets and Berglund, 1986; Fountain and Huizenga, 1996; ASHRAE, 1989). See also Höppe (1997) for discussion of other models. The Fanger model assumes a steady state of equilibrium has been reached whereas the Pierce Two-Node model considers transient conditions and thus allows the changes in body temperature with exposure time to be evaluated. Also, the Pierce Two-Node model breaks the body into two parts, an inner body and a skin layer with different temperatures in each. Both these models were developed initially to examine indoor comfort but their use has later been extended to outdoor conditions (Arens, Blyholder and Schiller 1984; Arens and Bosselman 1989; Arens, Gonzalez and Berglund, 1986; Bosselman, Arens, Dunbar and Wright, 1995). In the present discussion the Pierce Two-Node model will be focussed on to illustrate an approach to assessing thermal comfort. Other alternative models could also be used to achieve the same objectives.

Thermal comfort is a function of the body's core temperature, mean skin temperature, and skin wetness from unevaporated sweat. These variables are predicted by the Two-Node model as the outcomes of the following two heat balances. The rate of heat storage in the skin layer, S_{SK}, is given by the following equation

$$S_{SK} = Q_{CRSK} - E_{SK} - R_{SK} - C_{SK} \tag{10}$$

where Q_{CRSK} = rate of heat flow from core to skin,
E_{SK} = rate of evaporative heat loss from skin to surroundings,
R_{SK} = rate of radiative heat loss from skin to surroundings, and
C_{SK} = rate of convective heat loss from skin to surroundings.

The rate of heat storage in the body core is given by

$$S_C = M - W - E_{RES} - C_{RES} - Q_{CRSK} \tag{11}$$

where M = rate of metabolic heat production,
W = rate of mechanical work,
E_{RES} = rate of evaporative heat loss due to respiration, and
C_{RES} = rate of convective heat loss due to respiration.

The term Q_{CRSK} links these two equations.

The quantification of each of the terms in these equations is complex and it is not intended to enter into a detailed explanation here. A detailed description can be found in Doherty and Arens (1988) for example. It is clear, however, that various environmental factors will affect the particular terms. For example, E_{SK} depends on the relative humidity, air temperature and clothing level, while C_{SK} depends on air temperature, air velocity and clothing level. The radiative term R_{SK} will be strongly influenced by whether the person is in the sun or in shade. The metabolic and work rate terms, M and W depend on the activity level, as do the terms E_{RES} and C_{RES}.

Research is still in progress on how to apply these models to the problem of assessing the overall pedestrian comfort in an outdoor environment, see Appendix B. The next section presents an example of using the above model to assess the pedestrian comfort including the thermal effects.

5.2   An Example of Thermal Comfort Assessment

5.2.1   Basic Components and Criteria

For evaluating the combined impact of outdoor environment on pedestrian comfort, an example of a comprehensive model which considers three major components is that described by Soligo, Irwin and Williams (1993) and Soligo, Irwin, Williams and Schuyler (1997).

(1) **Wind Force** is assessed by the gust wind speed, and the Gust Equivalent Mean (GEM) described in Section 4.1. The acceptable ranges for three typical outdoor pedestrian activities, i.e. sitting, standing and walking, are assessed as described in Sections 4.1 and 4.2;

(2) **Thermal Comfort** is measured by the human body temperature using the Pierce Two-Node Model and the thermal sensation index (TSENS). The assessment of thermal comfort involves examining the heat balance of a human body to determine if excessive temperature change in the body is occurring after a certain exposure time to ambient conditions. Given a set of environmental and personal conditions and an exposure time, the model predicts core and skin compartment temperatures for the required exposure time. Then the TSENS scale is used to represent the level of pedestrian thermal comfort. A typical comfort range of TSENS can be set from -1.0 to +1.0, but the limits may be changed, depending upon the local climate for a specific project, to account for the effect of acclimatization of local residents.

(3) **Wind Chill** combines wind speed and air temperature to determine the chilling effect on the exposed skin. An equivalent temperature of -20°C (Equation 5 in Section 2.2), translating to a Wind Chill Index of 1166 kcal/m²h (equal to 1355 W/m²), is used as an onset of discomfort.

These three components are studied concurrently in the model. In order to pass the overall comfort in any given hour, all three individual comfort components must pass for that hour. The overall conditions are considered comfortable for a particular activity, if they pass all criteria for that activity at least 80% of the time.

5.2.2 Input Parameters and Results

Various input parameters are acquired from three primary sources, namely, a computerized sun/shade simulation, wind-tunnel tests, and meteorological records. While the wind-tunnel tests have been discussed in detail in previous sections, the sun/shade simulation and meteorological records need further explanation.

The presence or lack of direct sunlight experienced by a pedestrian can have a significant effect on thermal comfort. In order to determine if a particular location will experience sun or shade for various dates and times throughout the year, it is necessary to consider the angle of the sun and the geometry of the surrounding buildings. Building geometry can be modelled using three-dimensional computer graphics. The various times of year and times of day are represented in the computer model by varying the solar angles, making it possible to determine which sensors are exposed to sunlight or shade. This process is repeated for every hour of the day for each of the representative seasonal days. More details about the calculation of solar radiation are provided in Appendix A.

Meteorological data from the weather station closest to the study site typically contain hourly readings of wind direction and speed, ambient temperature, relative humidity, amount and opacity of cloud cover, precipitation and other information. The effective radiant field, the partial vapour pressure of water, and clothing insulation value are calculated or specified, and appended to each hourly meteorological data record for the use of the Two-Node Model. In addition, it is necessary to first predict the extent of clothing worn by the subject. Although, typically, a particular level of clothing will be associated with each season, day-to-day climate variations will produce day-to-day variations in clothing. Furthermore, in reality, the subject has the ability to take off or add clothing over the course of the day. One approach is to assume the person dresses based on an accurate weather forecast for the day and then to assume they can adjust by one clothing level either way to variations from the daily normal temperature.

The output lists the percentage of hours which passed in each category, for each sensor. An example of this output is presented in Table 5. The period of analysis and the hours of the day are also reported. Under each thermal category, the percentage passing is reported, and the percentage failing is split into two categories "cold" and "hot" which identifies the thermal sensation causing the failure. Specifically, Location 1 is identified as being thermally comfortable for sitting type activities 90.8% of the time. The remaining 9.2% is shown to be comprised of conditions being too cold 6.4% of the time and too hot 2.8% of the time. The overall rating is Comfortable for Standing at this location.

Table 5: Pedestrian Comfort Analysis Data Output

Afternoon Events (noon to 5 pm)

Period - of the year : 04 / 01 ---> 05 / 31
     - of the day : 12 : 00 ---> 17 : 00

Location	Wind Chill	Wind Force			Thermal			Overall				Category Rating
	Pass %	% of hrs. comfortable			% of hrs. comfortable			% of hrs. comfortable				
		Sitting	Standing	Walking	Sitting	Standing	Walking	Sitting	Standing	Walking	Severe	
					cold/hot	cold/hot	cold/hot					
1	100.0	56.8	85.9	96.9	90.8	93.1	94.6	51.7	80.2	91.6	000	standing
					6.4/2.8	3.6/3.3	1.1/4.3					
2	100.0	75.1	92.7	98.0	85.8	85.5	83.9	62.8	78.5	82.0	0.05	walking
					1.5/12.6	0.9/13.6	0.3/15.9					

The three comfort components are assessed individually as well as in combination, to determine an overall level of pedestrian comfort. For overall comfort, the three comfort components are combined into a single overall comfort evaluation for each of the three standard activities. In order to pass overall comfort in any given hour, all three individual comfort components and the overall assessment must pass for that hour. The purpose of assessing these three comfort components individually is to identify which component(s) cause a specific location to fail the overall pedestrian comfort criteria. By identifying the component(s), the type of mitigation required to improve conditions can be determined.

It is useful to compare the comfort conditions at various points around a building with reference points well away from the direct influence of the building. One reference point can be a sunny location and another a shady location.

6.   CONCLUDING REMARKS

The usefulness and attractiveness of outdoor areas near buildings are greatly affected by the local microclimate. The current state of the art of methods of systematically quantifying the microclimate are described in this document, as well as criteria for assessing human comfort. The massing of buildings and the design of the landscaping can have a significant impact on the comfort of outdoor areas, particularly in the way they influence wind flow patterns. Although some of these influences can be predicted from past experience and have been described in this document, in practice every building has its own idiosyncrasies. Therefore, whenever possible, it is best to evaluate the microclimate early in the design process using the techniques described herein.

The setting of criteria for comfort continues to be the subject of research. While there is a quite high degree of consensus on the safety criteria, there remains some diversity of opinion on the comfort criteria, as indicated by the variety of proposals summarised in Appendix C. The criteria described in Section 4.1 are thought to be a reasonable compromise between the conflicting demands of accuracy and simplicity based on current knowledge. However, further research, that draws on the subjective responses of users of outdoor spaces, would be very beneficial. The inclusion of detailed thermal effects, which entails greater complexity with many variables to contend with, is a relatively recent development and will benefit even more from additional research.

BIBLIOGRAPHY

Adrian, R.J., "Particle-Imaging Techniques for Experimental Fluid Mechanics", Annual Review of Fluid Mechanics, Volume 23, pages 261-304, 1991.

Anderson, S.R. "Humidex Calculation", Atmospheric Environment Service, CDS. No. 24-65, 1965.

Apperly, L.W. and Vickery, B.J., "The Prediction and Evaluation of the Ground Level Wind Environment", Proceedings of the Fifth Australasian Hydraulics and Fluid Mechanics Conference, University of Canterbury, Christchurch, New Zealand, pages 175-182, 1974.

Arens, E., Ballanti, D., Bennett, C., Guldman, S. and White, B., "Developing the San Francisco Wind Ordinance and Its Guidelines for Compliance", Building and Environment, Volume 24, Number 4, pages 297-303, 1989.

Arens, E. and Bosselmann, P., "Wind, Sun and Temperature - Predicting the Thermal Comfort of People in Outdoor Spaces", Building and Environment, Volume 24, Number 3, pages 315-320, 1989.

Arens, E., Gonzalez, R. and Berglund, L.G., "Thermal Comfort Under an Extended Range of Environmental Conditions", ASHRAE Transactions, Volume 92, Number 1, 1986.

Arens, E., Blyholder, A., and Schiller, G. "Predicting Thermal Comfort of People in Naturally Ventilated Buildings." ASHRAE Transactions, 90 (1), 1984.

Arens, E. and Peterka, J.A., "Control of Wind Climate Around Buildings", Journal of Transportation Engineering, American Society of Civil Engineers, Volume 110, pages 493-505, 1984.

Arens, E., "On Considering Pedestrian Winds During Building Design", Cambridge University Press, Proceedings of the International Workshop on Wind Tunnel Modeling Criteria and Techniques in Civil Engineering Applications, edited by Reinhold, T.A, pages 8-26, Maryland, USA, 1982.

ASCE, "Manual of Practice for Wind Tunnel Studies of Buildings and Structures", Manual Number 67, American Society of Civil Engineers, 1999.

ASHRAE, "Physiological Principals, Comfort and Health", Fundamentals Handbook, Chapter 8, 1989.

Aynsley, R.M., Melbourne, W. and Vickery, B.J., "Architectural Aerodynamics", Applied Sciences Publishers Ltd., London, 1977.

Aynsley, R.M., "Effects of Airflow on Human Comfort", Building Science, Volume 9, pages 91-94, 1974.

Baskaran, A., "Computer Simulation of 3D Turbulent Wind Effects on Buildings", PhD dissertation, Concordia University, Montreal, Canada, 1990.

Baskaran, A. and Stathopoulos, T., "Computational Evaluation of Wind Effects on Buildings", Building and Environment, Volume 24, Number 4, pages 325-333, 1989.

Benford, f. and Bock, J.E., "A Time Analysis of Sunshine", Transactions of the American Illumination Engineering Society, Volume 34, Number 200, 1939.

Beranek, W.J., "Wind Environment Around a Single Building of Rectangular Shape and Wind Environment Around Building Configurations", Institute for Building Materials and Building Structures, Heron Magazine, Volume 29, Number 1, pages 3-70, Netherlands, 1984.

Bluestein, M. and Zecher, J. "A New Approach to an Accurate Wind Chill Factor", Bulletin of the American Meteorological Society, Volume 80, Number 9, pages 1893-1899, 1999.

Borges, A.R.L. and Saraiva, J.A.G., "An Erosion Technique for Assessing Ground Level Winds", Proceedings of the Fifth International Conference on Wind Engineering, Fort Collins, Colorado, pages 235-242, 1979.

Borland, R., Owen, N., Hill, D. and Chapman, S., "Regulatory Innovations, Behaviour and Health: Implications of Research on Workplace Smoking Bans", International Review of Health Psychology, John Wiley and Sons, Volume 5, 1994.

Bosselman, P., Arens, E., Dunbar, K., and Wright, R., "Urban Form and Climate: Case Study, Toronto." Journal of the American Planning Assoc., Vol. 61, No. 2, pp. 226-239, 1995.

Bosselman, P. "Application of Computer Visualization and Mathematical Modeling Techniques in the Development of Planning Controls Designed to Protect Thermal Comfort for Pedestrians", Working Paper 549, Institute of Urban and Regional Development, University of California at Berkeley, Berkeley, CA, USA, 1991.

Bottema, M., "Wind Climate and Urban Geometry", PhD dissertation, Technische Universitiet Eindhoven, Facultyeit Bouwkunde, Vakgroep Fago, 1993.

Bottema, M., Leene, J.A. and Wisse, J.A., "Toward Forecasting of Wind Comfort", Journal of Wind Engineering and Industrial Aerodynamics, Volume 41-44, pages 2365-2376, 1992.

Bradbury, L.S.J., "Measurements with a Pulsed-Wire and Hot-Wire Anemometer in the Highly Turbulent Wake of a Normal Flat Plate", Journal of Fluid Mechanics, Volume 177, pages 473-497, 1976.

Buchhave, P., George, W.K. and Lumley, J.L., "The Measurement of Turbulence with the Laser-Doppler Anemometer", Annual Review of Fluid Mechanics, Volume 11, pages 443-503, 1979.

Bradshaw, P., "An Introduction to Turbulence and its Measurement", The Commonwealth Library of Science Technology Engineering and Liberal Studies, Pergamon Press, First Edition, 1971.

Carpenter, P., "Wind Speeds in City Streets - Full Scale Measurements and Comparison with Wind Tunnel Results", Proceedings of the Second Asia Pacific Symposium on Wind Engineering", Pergamon Press, June 1989.

Cheung, J.C.K., "High Wind Pressure on a Canopy or Podium Around the Base of a Tall Building", Proceedings of the Second Asia Pacific Symposium on Wind Engineering, pages 462-468, June 1989.

Cochran, L.S., Pielke, R.A. and Kovács, E., "Selected International Receptor-Based Air Quality Standards", Journal of Air and Waste Management Association, Volume 42, Number 12, December 1992.

Cochran, L.S. and Howell, J.F., "Wind Tunnel Studies for the Aerodynamic Shape of Sydney, Australia", Journal of Wind Engineering and Industrial Aerodynamics, Volume 36, pages 801-810, October 1990.

Cochran, L.S., "Full-Scale Ground Level Wind Study of the AMP Building, Brisbane", Baccalaureate Thesis, University of Queensland, Australia, 1979.

Cohen, H., Moss, S.M., and Zube, E.H., "Pedestrians and Wind in the Environment", Proceedings of the Tenth Conference of the Environmental Design Research Association, Boston, pages 71-82, 1979.

Cohen, H., McLaren, T.I., Moss, S.M., Petyk, R. and Zube, E.H., "Pedestrians and Wind in the Environment", Environment and Behavior Research Center, Institute for Man and Environment, Blaisdell House, University of Massachusetts/Amherst, Publication Number UMASS/IME/R-77/13, December 1977.

Comte-Bellot, G., "Hot-Wire Anemometry", Annual Review of Fluid Mechanics, Volume 8, pages 209-231, 1976.

Cook, N.J., "The Designers Guide to Wind Loading of Building Structures. Part 1 Background, Damage Survey, Wind data and Structural Classification. London, Building Research Establishment, and Butterworths, 1985.

Cooper, P.I., "The Absorption of Solar Radiation in Solar Stills", Solar Energy, Volume 12, Number 3, 1969.

Craze, D., "Linearized Anemometer Sensor Errors in Highly Turbulent Flows", University of Western Ontario, Boundary Layer Wind Tunnel Laboratory Report Number 1-73, June 1973.

Davenport, A.G., "An Approach to Human Comfort Criteria for Environmental Wind Conditions", Proceedings of the Colloquium on Building Climatology, Stockholm, Sweden, 1972.

Davenport, A.G., and Isyumov, N. "Application of the Boundary Layer Wind Tunnel to the Prediction of Wind Loading", International Research Seminar on Wind Effects on Buildings and Structures, Ottawa, Canada, 201-230, University of Toronto Press, 11-15 September 1967.

Deaves, D.M. and Harris, R.I., "A Mathematical Model of the Structure of Strong Winds", Construction Industry Research and Information Association (U.K.), Report #76, 1978.

Doherty, T.J. and Arens, E., "Evaluation of the Physiological Bases of Thermal Comfort Models", ASHRAE Transactions, Volume 94, Number 1, pages 1371-1385, 1988.

Driscoll, D.M. "Thermal Comfort Indexes: Current Uses and Abuses", National Weather Digest, Volume 17, Number 4, pages 33-38, 1992.

Dryden, L.H. and Kuethe, A.M., "The Measurement of Fluctuations of Air Speed by Hot-Wire Anemometry", National Advisory Committee on Aeronautics Technical Report 320, 1920.

Duffie, J.A. and Beckman, W.A. "Solar Engineering of Thermal Processes", John Wiley and Sons, Second Edition, 1991.

Durgin, F.H., "Evaluating Pedestrian Level Winds", Journal of Wind Engineering and Industrial Aerodynamics, submitted for publication, 2002.

Durgin, F.H., "Pedestrian Level Wind Criteria using the Equivalent Average" Journal of Wind Engineering and Industrial Aerodynamics, Volume 66, pages 215-226, 1997.

Durgin, F.H., "Equivalent Average Pedestrian Wind Comfort Criteria", Proceedings of the XIII American Society of Civil Engineers Structures Congress, pages 112-115, April 1995.

Durgin, F.H., "Pedestrian Level Wind Studies at the Wright Brothers Facility", Journal of Wind Engineering and Industrial Aerodynamics, Volume 41-44, pages 2253-2264, 1992.

Durgin, F.H., "Proposed Guidelines for Pedestrian Level Wind Studies in Boston: Comparison of Results from Twelve Studies", Building Environment, Volume 24, Number 4, pages 315-324, 1989.

Durgin, F.H., "Proposed Standards and a New Criteria for Pedestrian Level Wind Studies for Boston Massachusetts", Proceedings of the Fifth United States National Conference on Wind Engineering, Lubbock, Texas, 1985.

Durgin, F.H., "A Study of Pedestrian Level Winds for Battery Park City New York", Third United States National Conference of Wind Engineering, University of Florida, 1977.

Durgin, F.H. and Chok, A.W., "Pedestrian Level Winds: A Brief Review", American Society of Civil Engineers, Journal of the Structural Division, Number 108, pages 1751-1767, August 1982.

ESDU, Engineering Sciences Data Unit, "Strong Winds in the Atmospheric Boundary Layer . Pt 1: Mean Hourly Speeds", No. 82026. Issued September 1982 with amendments to 1993.

ESDU, Engineering Sciences Data Unit, "Strong Winds in the Atmospheric Boundary Layer . Pt 2: Discrete Gust Speeds", No. 83045. Issued November 1983 with amendments to 1993.

ESDU, Engineering Sciences Data Unit, "Characterization of Atmospheric Turbulence Near the Ground, Part II: Single Point Data for Strong Winds (Neutral Atmosphere)", Number 80520, Issued October 1985 with amendments to 1993.

Ettouney, S.M. and Fricke, F.R., "An Anemometer for Scale Model Environmental Wind Measurements", Building Science, Volume 10, pages 17-26, 1975.

Fanger, P.O. "Thermal Comfort", Danish Technical Press, Copenhagen, Denmark, 1972.

Fanger, P.O., Melikov, A.K., Hanzawa, H. and Ring, J., "Air Turbulence and Sensation of Drought", Energy and Buildings, Volume 12, Number 1, pages 21-39, 1988.

Fingerson, L.M. and Freymuth, P., "Thermal Anemometers", Fluid Mechanics Measurements, Hemisphere Publishing Corporation, Washington, pages 99-154, 1983.

Fountain, M., and Huizenga, C., "A Thermal Comfort Prediction Tool." ASHRAE Journal, Vol. 38, No. 9, pp. 39-42, 1996.

Gagge, A.P., Fobelets, A.P. and Berglund, L.G., "A Standard Predictive Index of Human Response to the Thermal Environment", ASHRAE Transactions, Volume 92, Number 2, pages 709-731, 1986.

Gagge, A.P., Stolwijk, J.A. and Hishi, Y., "An Effective Temperature Scale Based on a Simple Model of Human Physiological Regulatory Response", ASHRAE Transactions, Volume 77, Number 1, pages 247-262, 1971.

Gandemer, J., "Aerodynamic Studies of Built-Up Areas Made by CSTB at Nantes, France", Journal of Industrial Aerodynamics, Volume 3, pages 227-240, 1978.

Gandemer, J., "Wind Environment Around Buildings: Aerodynamic Concepts", Proceedings of the Fourth Conference on Wind Effects on Buildings and Structures, Cambridge, United Kingdom, pages 423-432, 1975.

Graber, A., Huhnergarth, A. and Schnotale, D., "Non-Rotating Instrument for Simultaneous Measurement of Wind Velocity and Wind Direction", Technisches Messen, Munich, Germany, pages 469-472, 1990.

Gerhardt, H.J. and Kramer, C., "Wind Comfort and Pollutant Transport in a Satellite City", Journal of Wind Engineering and Industrial Aerodynamics, Volume 41-44, pages 2343-2351, 1992.

Gerhardt, H.J. and Kramer, C., "Wind Shelters at the Entrance of Large Buildings - Case Studies", Journal of Wind Engineering and Industrial Aerodynamics, Volume 23, pages 297-3071, 1986.

Grant, R.H., Heisler, G.M. and Harrington, L.D., "Full-Scale Comparison of a Wind Tunnel Simulation of Windy Locations in an Urban Area", Journal of Wind Engineering and Industrial Aerodynamics, Volume 31, pages 335-341, 1988.

Grip, R.E., "An Investigation of the Erosion Technique for the Evaluation of Pedestrian Level Winds", Master's Thesis, Department of Civil Engineering, Massachusetts Institute of Technology, 1982.

Hanratty, T.J. and Campbell, J.A., "Measurement of Wall Shear Stress", Fluid Mechanics Measurements, Hemisphere Publishing Corporation, Washington, pages 559-615, 1983.

Heidorn, K.C., Murphy, M.C., and Davies, A.E., "Wind Tunnel Modelling of Roadways: Comparison with Mathematical Models." Proceedings, 82nd Annual Meeting, Air and Waste Management Association, Anaheim, California, 1989.

Höppe, P. "Aspects of Human Biometeorology in Past, Present and Future", International Journal of Biometeorology, 40: 19-23, 1997.

Hosker Jr., R.P., "Flow Around Isolated Structures and Building Clusters: A Review", ASHRAE Transactions, Volume 91, Number 2B, pages 1672-1692, 1985.

Howell, J.F., "Wind Issues at the Dandenong Taxation Building", Vipac Engineers and Scientists Pty. Ltd., Report for the Australian Taxation Office, 1987.

Huizenga, C., Zhang, H. and Arens, E., "A model of Human Physiology and Comfort for Assessing Complex Thermal Environments." Building and Environment, 2001.

Hunt, J.C.R., "Industrial and Environmental Fluid Mechanics", Annual Review of Fluid Mechanics, Volume 23, pages 1-41, 1991.

Hunt, J.C.R., Abell, C.J., Peterka, J.A. and Woo, H., "Kinematical Studies of the Flows Around Free or Surface Mounted Obstacles; Applying Topology to Flow Visualization", Journal of Fluid Mechanics, Cambridge University Press, Volume 86, pages 179-200, 1978.

Hunt, J.C.R., Poulton, E.C. and Mumford, J.C., "The Effects of Wind on People - New Criteria Based on Wind Tunnel Experiments", Building and Environment, Volume 11, pages 15-28, 1976.

Hunt, J.C.R., "Turbulent Velocities Near, and Fluctuating Surface Pressures on Structures in Turbulent Wind", Proceedings of the 4th International Conference on Wind Effects on Buildings and Structures, United Kingdom, 1975.

Irwin, H.P.A.H., "Instrumentation Consideration for Velocity Measurements", Proceedings of an International Workshop on Wind Tunnel Modelling Criteria and Techniques in Civil Engineering Applications, Cambridge University Press, pages 546-557, 1982.

Irwin, H.P.A.H., "A Simple Omnidirectional Sensor for Wind Tunnel Studies of Pedestrian Level Winds", Journal of Wind Engineering and Industrial Aerodynamics, Volume 7, pages 219-239, 1981.

Irwin, H.P.A.H., "A Simple Omnidirectional Sensor for Wind Tunnel Studies of Pedestrian Level Winds", Laboratory Technical Report #LTR-LA-242, National Aeronautical Establishment, National Research Council of Canada, 33 pages May 1980.

Irwin, P.A., Hunter, M.A., and Williams, C.J., "Microclimate Effects and Their Impact on Livability." Proceedings, Council on Tall Buildings and Urban Habitat Conference, Building for the 21st Century, London, Dec. 9-11, 2001.

Ishizaki, H., and Sung, I.W., "Influence of Adjacent Buildings to Wind", Proceedings of the Third Conference on Wind Effects on Buildings and Structures, Tokyo, Japan, pages 145-152, 1971.

Isyumov, N., "Full-Scale Studies of Pedestrian Winds: Comparison with Wind Tunnel and Evaluation of Human Comfort", Proceedings of the XIII American Society of Civil Engineers Structures Congress, pages 104-107, April 1995.

Isyumov, N., Helliwell, S., Rosen, S., and Lai, D., "Wind in Cities: Effects on Pedestrians and the Dispersion of Ground Level Pollutants", 6th Colloquium on Industrial Aerodynamics, Aachen, Germany, June 19-21, 1985.

Isyumov, N. and Davenport, A.G., "Comparison of Full Scale and Wind Tunnel Speed Measurements in the Commerce Court Plaza", Journal of Industrial Aerodynamics, Volume 1, pages 201-212, 1975(a).

Isyumov, N. and Davenport, A.G., "Ground Level Wind Environment in Built-Up Areas", Proceedings of the Fourth International Conference on Wind Effects on Buildings and Structures, Cambridge University Press, United Kingdom, pages 403-422, 1975(b).

Isyumov, N., "Studies of the Pedestrian Level Wind Environment at the Boundary Layer Wind Tunnel Laboratory of the University of Western Ontario", Journal of Industrial Aerodynamics, Volume 3, pages 187-200, 1978.

Jackson, P.S., "The Evaluation of Windy Environments", Building Environment, Volume 13, pages 251-260, 1978.

Jamieson, N.J., Carpenter, P. and Cenek, P.D., "The Effect of Architectural Detailing on Pedestrian-Level Wind Speeds", Journal of Wind Engineering and Industrial Aerodynamics, Volume 41-44, pages 2301-2312, 1992.

Kamei, I. and Maruta, E., "Study on Wind Environmental Problems Caused Around Buildings in Japan", Journal of Industrial Aerodynamics, Volume 4, pages 307-331, 1979.

King, C.V., "On the Convection of Heat from Small Cylinders in a Stream of Fluid", Philosophical Transactions of the Royal Society, London, Volume A214, page 373, 1914.

Lawson, T.V., "The Wind Content of the Built Environment", Journal of Industrial Aerodynamics, Volume 3, pages 93-105, 1978.

Lawson, T.V. and Penwarden, A.D., "The Effects of Wind on People in the Vicinity of Buildings", Proceedings of the Fourth International Conference on Wind Effects on Buildings and Structures, Cambridge University Press, United Kingdom, pages 605-622, 1975.

Lee, B.E. and Hussain, M., "The Ground Level Wind Environment Around the Sheffield University Arts Tower", Journal of Industrial Aerodynamics, Volume 4, pages 334-341, 1979.

Letchford, C.W. and Ginger, J.D., "Wind Environment Assessment: A Case Study in the Brisbane CBD", Proceedings of the Third Asia Pacific Symposium on Wind Engineering, Volume 2, pages 973-978, Hong Kong, December 1993.

Letchford, C.W. and Isaacs, L.T., "Full Scale Measurement of Wind Speeds in an Inner City", Journal of Wind Engineering and Industrial Aerodynamics, Volume 44, pages 2331-2341, 1992.

Livesey, F., Morrish, D., Mikitiuk, M., Isyumov, N. and Davenport, A.G., "Enhanced Scour Tests to Evaluate Pedestrian Level Winds", Journal of Wind Engineering and Industrial Aerodynamics, Volume 41-44, pages 2265-2276, 1992.

Livesey, F., Inculet, D., Isyumov, N. and Davenport, A.G., "A Scour Technique for the Evaluation Pedestrian Winds", Journal of Wind Engineering and Industrial Aerodynamics, Volume 36, pages 779-789, 1990.

Lohmeyer, A., Fassirinner, H., Schmitt, H. and Fehrenbach, K., "Case Study: Quantitative Determination of Pedestrian Comfort Near a High-Rise Building", Energy and Buildings, Volume 11, pages 149-156, 1988.

Masterton, J.M., and Richardson, F.A.. "A Method of Quantifying Human Discomfort due to Excessive Heat and Humidity", Environment Canada Report CLI 1-79, Downsview, Ontario, 1979.

Melbourne, W.H., "Criteria for Environmental Wind Conditions", Journal of Industrial Aerodynamics, Volume 3, pages 241-249, 1978.

Melbourne, W.H. and Joubert, P.N., "Problems of Wind Flow at the Base of Tall Buildings", Proceedings of the Third International Conference on Wind Effects on Buildings and Structures, Tokyo, pages 105-114, 1971.

Murakami, S., Iwasa, Y. and Morikawa, Y., "Study on Acceptable Criteria for Assessing Wind Environment on Ground Level Based on Resident's Diaries", Journal of Wind Engineering and Industrial Aerodynamics, Volume 24, pages 1-18, 1986.

Murakami, S., Fuji, and Kunio, "Turbulence Characteristics of Wind Flow at Ground Level in Built-Up Areas", Journal of Wind Engineering and Industrial Aerodynamics, Volume 15, pages 133-144, 1984.

Murakami, S., "Wind Tunnel Modeling Applied to Pedestrian Comfort", Proceedings of the International Workshop on Wind Tunnel Modeling Criteria and Techniques in Civil Engineering Applications, edited by Reinhold, T.A, pages 468-503, Maryland, USA, 1982.

Murakami, S. and Deguchi, K., "New Criteria for Wind Effects on Pedestrians", Journal of Wind Engineering and Industrial Aerodynamics, Volume 7, pages 289-309, 1981.

Murakami, S., Uehara, K. and Deguchi, K., "Wind Effects on Pedestrians: New Criteria Based on Outdoor Observation of Over 2000 Persons", Proceedings of the Fifth International Conference on Wind Engineering, Fort Collins, Colorado, pages 277-288, July 1979.

Murakami, S., Uehara, K. and Komine, H., "Amplification of Wind Speed at Ground Level Due to Construction of High Rise Building in an Urban Area", Journal of Industrial Aerodynamics, Volume 4, pages 343-370, 1979.

Nagib, H.M. and Corke, T.C., "Wind Microclimate Around Buildings: Characteristics and Control", Journal of Wind Engineering and Industrial Aerodynamics, Volume 16, pages 1-15, 1984.

Oke, T.R., "Street Design and Urban Canopy Layer Climate", Energy and Buildings, Volume 11, pages 103-111, 1988.

Olsen, E. and Strindehag, O., "An Investigation of a Sonic Flowmeter", Journal of Wind Engineering and Industrial Aerodynamics, Volume 37, pages 245-253, 1991.

Openspace, "Preserving Sunlight and Open Air in New York City", Volume 2 Number 3, 1988.

Owen, N. and Halford, K., "Psychology, Public Health and Cigarette Smoking", Australian Psychologist, Volume 23, Number 2, July 1988.

Penwarden, A.D., Grigg, P.F. and Rayment, R., "Measurement of Wind Drag on People Standing in a Wind Tunnel", Building and Environment, Volume 13, pages 75-84, 1978.

Penwarden, A.D. and Wise, A.F.E., "Wind Environment Around Buildings", Building Research Establishment Digest, Department of the Environment, Her Majesty's Stationery Office, United Kingdom, Pages 1-52, 1975.

Penwarden, A.D., "Acceptable Wind Speeds in Towns", Building Science, Volume 8, pages 259-267, 1973.

Perry, A.E.," Hot-Wire Anemometry", Oxford University Press, New York, 1982.

Peterka, J.A., Meroney, R.N. and Kothari, K.M., "Wind Flow Patterns About Buildings", Journal of Wind Engineering and Industrial Aerodynamics, Volume 21, pages 21-38, 1985.

Ratcliff, M.A. and Peterka, J.A., "Comparison of Pedestrian Wind Acceptability Criteria", Journal of Wind Engineering and Industrial Aerodynamics, Volume 36, pages 791-800, 1990.

Reed, D.A. "Expert Systems in Wind Engineering." Journal of Wind Engineering and Industrial Aerodynamics, 33, pages 487-494, 1990.

Rush, R., "Technics: Structuring Tall Buildings", Progressive Architecture, Reinhold Publishing Company Inc., December 1980.

Sadeh, W.Z., "Trends in Turbulence Measurements", Proceedings of the Third International Symposium on Refined Flow Modelling and Turbulence Measurements, Tokyo, Japan, July 1988.

Sandborn, V.A., "Class Notes for Experimental Methods in Fluid Mechanics", Department of Civil Engineering, Colorado State University, Fourth Edition, 1981.

Sandborn, V.A., "Resistance Temperature Transducers", Metrology Press, Fort Collins, Colorado, 1972.

Sasaki, R., Uematsu, Y., Yamada, M. and Saeki, H. "Application of Infrared Thermography and Knowledge-Based System to the Calculation of Pedestrian-Level Wind Environment around Buildings." Second International Symposium on Computational Wind Engineering, CWE'96, Colorado State University, Fort Collins, CO, USA, August 4-8, 1996.

Schubauer, G.B. and Klebanoff, P.S., "Theory and Application of Hot Wire Instruments in the Investigation of Turbulent Boundary Layers", N.A.C.A. Wartime Report, Advance Confidential Report 5K27 W86, 1946.

Simiu, E. and Scanlan, R.H., "Wind Effects on Structures: an Introduction to Wind Engineering", Third Edition, John Wiley & Sons, New York, 590 pages, 2000.

Siple, P.A., and Passel, C.F., "Measurements of Dry Atmospheric Cooling in Subfreezing Temperatures, Proceedings of the American Philosophical Society 89:177, 1945.

Soligo, M.J., Irwin, P.A., Williams, C.J., Schuyler, G.D."A Comprehensive Assessment of Pedestrian Comfort Including Thermal Comfort", Eighth US National Conference on Wind Engineering, Baltimore, MD, US, June 1997.

Soligo, M.J., Irwin, P.A., Williams, C.J., Schuyler, G.D., "Pedestrian Comfort: A Discussion of the Components to Conduct a Comprehensive Assessment", Proceedings of the XIII American Society of Civil Engineers Structures Congress, pages 108-111, April 1995.

Soligo, M.J., Irwin, P.A. and Williams, C.J., "Pedestrian Comfort Including Wind and Thermal Effects", Proceedings of the Third Asia Pacific Symposium on Wind Engineering, Hong Kong, 1993.

Sparks, P.R. and Elzebda, "A Comparison of Pedestrian Comfort Criteria Applied to a City Center", Journal of Wind Engineering and Industrial Aerodynamics, Volume 15, pages 123-132, 1983.

Stathopoulos, T. "Computational Wind Engineering: Past Achievements and Future Challenges." Keynote Lecture at the Second International Symposium on Computational Wind Engineering, CWE'96, Colorado State University, Fort Collins, CO, USA, August 4-8, 1996.

Stathopoulos, T. and Baskaran, A. "Computer Simulation of Wind Environmental Conditions around Buildings." Engineering Structures, 18(11), 876-885, 1996.

Stathopoulos, T. and Saathoff, P., "Pedestrian Wind Environment Criteria for the City of Montreal", A report prepared for the City of Montreal, Centre for Building Studies, Concordia University, Montreal, Canada, 1989.

Stathopoulos, T. and Wu, H., "Generic Models for Pedestrian-Level Winds in Built-Up Regions", Journal of Wind Engineering and Industrial Aerodynamics, Volume 54-55, pages 515-525, 1994.

Stathopoulos, T., Wu, H. and Bedard, C., "Wind Environment Around Buildings: A Knowledge Based Approach", Journal of Wind Engineering and Industrial Aerodynamics, Volume 41-44, pages 2377-2388, 1992.

Steadman, R.G. "Indices of Windchill of Clothed Persons", Journal of Applied Meteorology, Volume 18, pages 861-885, 1971.

Taylor, G.I., "Skin Friction of the Wind on the Earth's Surface", Proceedings of the Royal Society, Volume 92, 1916.

Uematsu, Y., Yamada, M., Higashiyama, H. and Orimo, T., "Effects of the Corner Shape of High-Rise Buildings on the Pedestrian-Level Wind Environment with Consideration for Mean and Fluctuating Wind Speeds", Journal of Wind Engineering and Industrial Aerodynamics, Volume 41-44, pages 2277-2288, 1992.

Uematsu, Y. and Yamada, M., "Application of Infrared Thermography to the Evaluation of Pedestrian-Level Winds Around Buildings", International Conference on Experimental Fluid Mechanics, Chinese Aerodynamic Research Society, Beijing, China, 1991.

Van der Hoven, "Power Spectrum of Horizontal Wind Speed in the Frequency Range from 0.0007 to 900 Cycles per Hour", Journal of Meteorology, Volume 14, pages 160-164, 1957.

Ville de Montreal, "Cadre Reglementaire: Arrondissement Ville-Marie", Module de la gestion du Developpement, Service de l'habitation et du developpement urbain, 1992.

Visser, G. Th., "Windhindercriteria: Een Literatuuronderzoek Naar en Voorstellen voor het Hanteren Van Uniforme TNO-Windhindercriteria (in Dutch)", Report Number 80-02746, IMET-TNO, Apeldoorn, Netherlands, 1980.

Wellington City Ordinances, Victoria University of Wellington, Wellington, New Zealand Works Central Laboratories Report 88-B9412.

Williams, C.J., Soligo, M.J. and Cote, J., "A Discussion of the Components for a Comprehensive Pedestrian Level Comfort Criteria", Journal of Wind Engineering and Industrial Aerodynamics, Volume 41-44, pages 2389-2390, 1992.

Williams, C.J., Hunter, M.A. and Waechter, W.F., "Criteria for Assessing the Pedestrian Wind Environment", Journal of Wind Engineering and Industrial Aerodynamics, Volume 36, pages 811-815, 1990.

Williams, C.D. and Wardlaw, R.L., "Determination of the Pedestrian Wind Environment in the City of Ottawa Using Wind Tunnel and Field Measurements", Journal of Wind Engineering and Industrial Aerodynamics, Volume 41-44, pages 2253-2264, 1992.

Wise, A.F.E., Sexton, D.E. and Lillywhite, M.S., "Air Flow Around Buildings", Proceedings of the Urban Planning Research Symposium, Building Research Station, London, England, pages 71-91, 1965.

Wu, H., "Pedestrian-Level Wind Environment Around Buildings", Ph.D dissertation, Concordia University, Montreal, Canada, 1994.

Wu, H. and Stathopoulos, T., "Computer-Aided Prediction of Pedestrian Level Wind Environment Around Buildings", Proceedings of the Third Canadian Conference on Computing in Civil and Building Engineering Systems, Montreal, 1996.

Wu, H., Stathopoulos, T. and Bedard, C., "A Knowledge-Based System for Predicting and Improving Pedestrian Wind Conditions", Civil Engineering Systems, Volume 12, pages 191-205, 1995.

Wu, H. and Stathopoulos, T., "Further Experiments on Irwin's Surface Wind Speed Sensor", Journal of Wind Engineering and Industrial Aerodynamics, Volume 53, pages 441-452, 1994.

Wu, H., and Stathopoulos, T. "Infrared-Thermography Technique for Pedestrian Wind Evaluation." Proceedings of Third Asia-Pacific Symposium on Wind Engineering, Hong Kong, 967-972, December 1993(a).

Wu, H. and Stathopoulos, T., "Wind-Tunnel Techniques for the Assessment of Pedestrian-Level Winds", American Society of Civil Engineers Journal of Engineering Mechanics, Volume 119, Number 10, pages 1920-1936, October 1993(b).

APPENDIX A

CALCULATION OF SOLAR RADIATION

To establish the comfort level at a given location the condition of sun or shade needs to be estimated as a function of time of day and day of year. The sun or shade dichotomy is best defined by knowledge about the direction of beam radiation, and so the initial focus here will be on creating shaded and sunlit areas around the buildings of interest.

The supply of energy from the sun, well above the atmosphere, is relatively constant with time except for small variations (about ±3% in Equation A1) due to the variation in the earth-sun distance. The World Radiation Center (Duffie and Beckman, 1991) has defined a Solar Constant (G_{sc}) of 1367 W/m^2 as being the power from the sun per unit area, outside the atmosphere, perpendicular to the direction of propagation of the radiation at the mean earth-sun distance of 1.495×10^{11} m. On any given day of the year (n) this power supply impinging on a plane normal to direction of the radiation (G_{on}) outside the atmosphere is given by Equation A1. Note that n is the sequential day of the year (i.e. January 01 is $n = 1$) and that all angles are in degrees.

$$G_{on} = G_{sc}\left[1 + 0.033\cos\left(\frac{360n}{365}\right)\right], \qquad 1 \leq n \leq 365 \tag{A1}$$

This available energy is diminished significantly by the atmosphere on its way to a terrestrial pedestrian location and is also very dependent on the time of day being considered.

The position of the sun during the day is calculated using the concept of "solar time" which is based on the apparent motion of the sun across the sky. Solar noon is the time when the sun crosses the meridian of the observer. The difference between solar time and standard time is caused by two principal mechanisms (Equations A2 to A4) and is calculated in minutes. The first correction is due to the longitude of the observer (L_{loc}) relative to the meridian in which the local standard time is based (L_{st}). The observation that it takes four minutes for the sun to pass over one degree of longitude completes the first component of the correction in Equation A2.

The second correction in the equation of time is due to the perturbations in the earth's rate of rotation (E) described in Equations A3 and A4. The two corrections used to convert standard time to solar time are given in minutes and do not include the one-hour variation if daylight saving time is in effect. For more discussion of these issues see Duffie and Beckman (1991).

$$\text{(solar time)} - \text{(standard time)} = 4(L_{st} - L_{loc}) + E \tag{A2}$$

where,

$$E = 0.01719 + 0.42815\cos(B) - 7.35205\sin(B) - 3.34976\cos(2B)$$
$$- 9.37199\sin(2B)$$
(A3)

$$B = (n-1)\left[\frac{360}{365}\right], \qquad 1 \leq n \leq 365$$
(A4)

Once the solar time has been established the direction of incoming radiation (and hence the shadow layout) may be calculated from some well established relationships (Benford and Bock, 1939; Cooper, 1969). Defining the parameters of interest:

φ **Latitude**, the angular location north or south of the equator, positive north.

δ **Declination**, the angular position of the sun at solar noon with respect to the plane of the equator, north positive.

$$\delta = 23.45\sin\left(360\frac{284+n}{365}\right)$$
(A5)

β **Slope**, the angle between the plane of the surface in question and the horizontal. The horizontal ground in a pedestrian study gives $\beta = 0°$.

γ **Surface azimuth angle**, the deviation of the projection on a horizontal plane of the normal to the surface from the local meridian, with zero due south, east negative and west positive.

ω **Hour angle**, the angular displacement of the sun east or west of the local meridian due to rotation of the earth on its axis at 15° per hour; morning negative and afternoon positive.

θ **Angle of incidence**, the angle between the beam radiation on a surface and the normal to that surface (the normal is vertical for horizontal ground).

$$\cos\theta = \sin\delta \sin\varphi \cos\beta - \sin\delta \cos\varphi \sin\beta \cos\gamma + \cos\delta \cos\varphi \cos\beta \cos\omega$$
$$+ \cos\delta \sin\varphi \sin\beta \cos\gamma \cos\omega + \cos\delta \sin\beta \sin\gamma \sin\omega$$
(A6)

By way of example, calculate the angle of incidence for beam radiation at Denver, Colorado ($\varphi = 39°N$, $L_{loc} = 104.9°W$ and $L_{st} = 105.0°W$) at 10:30 am on February 13 (ie $n = 44$ and $B = 42°$). The pedestrian location is an open horizontal plaza (ie $\beta = 0°$). Using Equations A2 to A4 to find the solar time:

(solar time) = (standard time) + 4(105.0 - 104.9) - 14.3 = (standard time) - 13.9

Thus 10:30 am mountain standard time is actually 10:16 am solar time. With this time known, the hour angle is given below.

$$\omega = -\left(1 + \frac{44}{60}\right)15° = -26°$$

From Equation A5, $\delta = -14°$. Using Equation A6, and noting that three terms drop out with $\sin(\beta)=0$.

$$\cos\theta = -0.152 + 0.678 = 0.526 \qquad \therefore \quad \theta = 58°$$

With the incoming radiation vector established for any given time, the location of the shaded or sunny areas around the site may be obtained as a function of time of day. The direct radiation dominates the comfort level in colder climates. However, in tropical locations the ambient local temperature (contributed to by diffuse radiation as well) may also be an important parameter in the prediction of comfort. In either case, the power of the radiation needs to be assessed for the chosen city on an hourly basis using tables of solar insolation. The direct solar insolation impacting pedestrian comfort is reduced from the extraterrestrial value in Equation A1 by processes such as scattering from dust and particulates (generally limited to wavelengths $\lambda < 0.6$ μm), and absorption of some spectral energies by various molecules. Consequently, local solar and cloud cover data from a nearby meteorological station (usually hourly or daily energy inputs in kJ/m² are available) should be used in the calculation of pedestrian comfort, rather than an approximate value derived from typical atmospheric conditions for that latitude and time of year.

ADDITIONAL INFORMATION ON INSTRUMENTATION AND
COMPUTATIONAL TECHNIQUES

POINT METHODS

Laser-Doppler Anemometry The Laser-Doppler Anemometer (LDA) measures fluid velocities by detecting the Doppler frequency shift of laser light that has been scattered by small particles moving with the fluid (Buchhave et al, 1979). This non-intrusive measurement is independent of thermophysical properties of fluids. The relation between the wind velocity and the frequency shift is linear and needs no calibration. The LDA is capable of sensing one or more velocity components and flow reversals, but a separate lens, thus additional alignment, is needed for 3-D measurements. It also measures very low velocities adequately because it is not influenced by the "free convection" effects that affect thermal anemometer readings.

Although LDA measurements can be made from the naturally existing particles in the flow in water flumes, artificial seeding is normally used in wind tunnels by adding particles of known characteristics. The seeding particles have to be small enough to follow the medium and could be generated by different types of smoke generators, humidifiers, etc. The LDA system is expensive, somewhat difficult to operate and generally unsuitable for routine tests. As a result, the Laser-Doppler technique has not seen extensive use in the context of pedestrian level wind studies. However, recent advances in fibre optic elements, signal processors and computer software may in the future make this technique more "user friendly".

Particle-image Velocimetry In order to measure instantaneous flow fields as a whole, a class of new techniques have been developed in the field of experimental fluid mechanics. For instance, following conventional flow visualization techniques, particle-image velocimetry (Adrian, 1991) uses low-density particles as tracking markers. During the measurement, markers in the fluid are illuminated by a sheet of pulsed light. The motion of marked regions is recorded by a video camera at two or more times, and the local velocity is then estimated from the displacement of a marker over a short time interval. Particle-image velocimetry is now capable of providing accurate measurements in two dimensions for a variety of laboratory-scale flows in a wide speed range. It is particularly powerful for measuring the turbulent flow field around complex geometrical configurations.

The obvious advantage in this group of techniques is the direct, accurate, and instantaneous measurement of velocity vectors for the whole flow field. At present, these techniques require the application of sophisticated mathematical theory and appropriate computer software and hardware. This has been an obstacle to their routine use in pedestrian level wind studies but with advances in computer techniques for image recording and data analysis this might change in future.

Other Instrumentation The wind induces on a human body an aerodynamic force, which directly affects human activities. This force has been evaluated by three similar simple methodologies developed specifically for this purpose, namely, spherical flow indicator (Isyumov and Davenport, 1975b), cylindrical force indicator (Ettouney and Fricke, 1975) and optical dynamometer (Beranek, 1984). The results obtained from these types of instrument are representative of the average wind force and direction on the human body represented by a cylinder or ball. However, their application is limited due to their low frequency response, non-linearity in the calibration and relatively large size as well as difficulties in data processing and interpretation. Thus, they have not been adopted for routine use.

Other instruments used in the application of point methods can also be found in the literature. These include the deflection velocimeter (Gräßer et al, 1990), the sonic flowmeter (Olsen and Strindehag, 1991), and devices to measure surface shear stresses such as the Stanton tube, the sublayer fence, and the electrochemical technique (Hanratty and Campbell, 1983). Note that the widely used Irwin sensor essentially is a surface shear stress technique but it has the distinct advantage of not requiring any alignment with the flow.

AREA METHODS

Erosion Test The erosion test identifies the windy zones by scour patterns in particles distributed over the model (Borges and Saraiva, 1979, and Livesey et al, 1990). Briefly, non-cohesive particles, like dry sand and cereals, are uniformly sprinkled to a few grains deep on the model site so that the same threshold wind speed for the particle movement can be applied to the entire field. Prior to a model test, the mean speed (U_n) for which the material starts moving is recorded at a reference height - typically the gradient height. After the building model is placed into the tunnel, the wind speed is increased from zero in several successive steps. If the loose material is blown away from certain zones around buildings when the mean wind speed at the reference height is U_r, the ratio of U_n/U_r can be used as an indicator of building-induced speed increases at the ground level.

The greatest advantages of this technique are its continuous coverage of large areas and its ability to provide visualization of the flow field. Quantitative results can be derived by analyzing the scour patterns with respect to the threshold speed. If a video record is played throughout the test, wind directions are possible to be identified. Digital image processing techniques have made the erosion technique more applicable as a routine method (Livesey et al, 1992).

However, what the erosion test really measures cannot be fully determined; the scour patterns do not show specifically the mean wind speed nor the turbulence intensity. Other practical difficulties still remain with this technique; for instance, how to select the size and geometry of the material grains, how to spread the material uniformly and how to collect it after each step of measurements. Therefore, its use tends to be limited to early qualitative studies of projects aimed at identifying obvious accelerated flow zones and not for detailed of final studies providing quantitative information.

Surface Flow Visualization Surface flow visualization in boundary-layer wind tunnels follows principles similar to those of the erosion technique. However, the visualization material and the operation procedures are different. Used in the surface visualization are fluid mixtures, like paraffin oil with kaolin powder, pigment paint and other materials. The test is normally conducted under a selected low wind speed and the wind-tunnel running time varies with the material used. The mixture is moved away from its original position by wind force and provides a qualitative indication of wind flow. Wind directions, in an average sense, may be inferred from the visualized flow field. Moreover, the interaction of wind flows around a group of buildings can be investigated and understood with the help of visualized flow patterns.

Surface flow visualization is easier to perform than the erosion test. It is capable of providing a high resolution picture of the flow patterns on the wind-tunnel floor. This type of visual presentation again helps lay persons in aerodynamics, such as architects and city planners more readily understand the wind flow patterns influencing the comfort or safety of a particular area. Other visualization techniques, like smoke streaklines emitted from holes in the model, particle injection, tufts and directional vanes (point methods), can also be a similar type of visual aid.

Infrared Thermography Infrared thermography has been applied to pedestrian-level wind studies (Uematsu and Yamada, 1991, and Wu and Stathopoulos, 1993a). The methodology is based on the fact that the heat transfer from a heated body to the flow is closely related to the flow conditions near the body surface. A simple experimental set-up consists of a heated thick acrylic plate on which the building model is placed, and an infrared camera that detects the temperature distribution on the floor plate and registers it by colour pictures or video records. Research into the functional relationship between the temperature pattern and the flow conditions has been reported by Wu and Stathopoulos, 1993a.

The advantage of this technique in comparison to other area methods is that no extra materials are introduced in the measurement field. The methodology is potentially capable of providing a quantitative evaluation of wind flows through digital image processing techniques. Only one wind speed, instead of a number of speeds as in the erosion test, need be tested. The temperature difference between the measurement plate and the air flow can be set at a relatively low value - say 10 degrees C - so that alteration of the wind flow by heat transfer from the heated plate becomes negligible. Although still under development, the infrared thermography technique for the evaluation of pedestrian-level winds does show potential.

KNOWLEDGE-BASED COMPUTER SYSTEMS

There are many building projects where some knowledge the building's influence on the wind flows around it and on methods of improving them would be useful in the early stages of design. Precise knowledge may not be important at this stage. Also, there are smaller projects where detailed studies would not be warranted for cost and timing reasons. In these cases wind consultants may be able to provide an approximate assessment based on experience without undertaking a special wind-tunnel study. Some research has been undertaken to examine the possibility of developing knowledge-based expert computer systems that could take on this role.

For example, to predict thermal comfort of people in outdoor places, Bosselmann (1991) combined wind-tunnel studies, mathematical modelling of human thermal regulatories, sun access computer modules, 3-D databases of open spaces and buildings, and other relevant knowledge into a new procedure for urban design studies for the cities of Toronto, San Francisco and New York. Bottema (1993) represented his findings from wind-tunnel experiments by rules of thumb for building design and city planning in a computerized environment. Similar work was also reported by Sasaki et al (1996).

Stathopoulos et al (1992) and Wu and Stathopoulos (1996) reported research where knowledge on wind flows around buildings was collected systematically from literature and generalized by further analysis and supplementary experiments. They then developed a computer system consisting of a series of databases containing information on aerodynamics from building models, terrain topography, weather data, discomfort wind criteria, city guidelines, multi-objective decision modules for remedy selection and others. These databases were integrated with computer graphics to serve as a consultation tool for building design and city planning, at least at the preliminary stage.

The advantages of knowledge-based approaches can be summarized as follows: (1) No special knowledge is required for running the system and complex domain knowledge becomes accessible to building designers and city planners via a user-friendly interface. (2) A usually time-consuming consultation process can be executed consistently in a shorter time period. (3) The evaluation of different design alternatives and potential remedies at early design stages is made more feasible. (4) The system can be expanded as more information of pedestrian-level winds becomes available, since in a knowledge-based system the domain knowledge is explicit and separate from the inference mechanism of the system.

COMPUTATIONAL WIND ENGINEERING (CWE)

Computational Wind Engineering (CWE) is a new branch of Wind Engineering dealing with the application of Computational Fluid Dynamics (CFD) methods to wind engineering and building aerodynamics problems. These include the evaluation of wind-induced pressures on buildings and other structures, as well as the wind velocity field around buildings, which is necessary for the evaluation of dispersion of pollutants around buildings and the assessment of wind environmental conditions at the pedestrian level, see Stathopoulos (1996).

Bottema (1993) has attempted the evaluation of pedestrian level wind conditions in the vicinity of an isolated building by using the CWE approach and a simple k-ε turbulence model but with only limited success. The feasibility of CWE applications in the area of wind environmental conditions at the pedestrian level has also been examined by Stathopoulos and Baskaran (1996) for a specific case study. The comparisons of computed results with those from wind tunnel tests indicated the computed results were typically within 30% of the wind tunnel data.

At the present time CWE is a rapidly developing field. The ability to produce graphic depictions of flow fields is a very attractive feature and with further development and verification it can be expected to take on an expanding role in the future.

THERMAL COMFORT MODELLING

The Two-Node Model originally developed by Pharo Gagge at the J.P. Pierce Foundation has been made available in easily useable Windows format by Fountain and Huizenga (1996) and is available from ASHRAE, Inc.'s publications department in Atlanta as the ASHRAE Thermal Comfort Model.

For all its thermophysiological detail, the Two-Node Model is a severely simplified approximation of the human body. The body is treated as a 0.6 m diameter sphere, with clothing uniformly distributed across its surface. New multi-segment models have been developed that allow the thermal effects of environmental asymmetries (wind, radiation, and temperature) to be explicitly considered for each body segment (Huizenga et al, 2001). They allow the clothed and unclothed areas of the body to be distinguished, which is very important for outdoor exposures to sun and wind. They have been spurred by the automobile industry, in that automobile heating and cooling systems must respond to the strongly asymmetrical environments and transient exposures often experienced by people as they enter and occupy cars. The new models are well tested for their thermophysiological predictions, but they are at early stages in predicting the subjective comfort consequences of people's thermal states. Research on this is being sponsored by the U.S. Department of Energy's National Renewable Energy Laboratory. When available, such models will make it possible for designers to simulate transients of environmental factors in time, either as the environmental factors change in a given space, or as a pedestrian traverses an area containing differing microclimates.

APPENDIX C

**SUMMARY OF VARIOUS WIND CRITERIA FOR
PEDESTRIAN COMFORT AND SAFETY**

C.1   Introduction

Over the years, a number of wind criteria have been developed for assessing pedestrian comfort and safety, based on field observations and wind-tunnel experiments. Selected criteria are summarized here in their original formats together with some brief explanations and notes. Several reviews and comparisons of criteria have been published. Examples are Melbourne (1978), Ratcliff and Peterka (1990) and Durgin (2002). Melbourne (1978) was of the opinion that the level of consistency between the criteria was quite high. However, Ratcliff and Peterka (1990) found sufficient differences to recommend considering several different sets of criteria and then forming a judgement based on overall results. Durgin (2002) focused on optimizing the probability level at which the criteria are set as discussed further below.

C.2   Penwarden (1973)

This author defined mean wind speeds for the onset of discomfort, unpleasant conditions and dangerous conditions. No particular probabilities of occurrence were associated with these speeds.

Mean Speed (m/s)	Perception
5	Onset of discomfort
10	Definitely unpleasant
20	Dangerous

C.3   Davenport (1972), Isyumov and Davenport (1975b)

The wind speeds were defined in terms of Beaufort numbers and the comfort criteria depended on the type of activity expected at each location. Also, the criteria were expressed in terms of frequency of occurrence. They are summarized in the following table.

Activity	Areas applicable	RELATIVE COMFORT			
		Perceptible	Tolerable	Unpleasant	Dangerous
1. Walking fast	Sidewalks	5	6	7	8
2. Strolling, skating	Parks, entrances, skating rinks	4	5	6	8
3. Standing, sitting - short exposure	Parks, plazas	3	4	5	8
4. Standing, sitting - long exposure	Outdoor restaurants bandshells, theatres	2	3	4	8
Representative criteria for acceptability			<1 occn/wk	<1 occn/mo	<1 occn/yr

The Beaufort numbers in the table were for temperatures above 10°C. At lower temperatures, relative comfort level were reduced by one Beaufort number for every 20°C reduction in temperature. Also, since Beaufort numbers are associated with wind speed ranges, rather than specific speeds, it was necessary, when computing frequencies of occurrence, to define a mean speed for each Beaufort number. The following speeds were used.

Beaufort Number		0	1	2	3	4	5	6	7	8
Mean wind speed	m/s mph	0	0.9	2.45	4.41	6.7	9.3	12	16	18.9

The speeds in the table are mean values at 10 m height in open terrain. To compensate for the different velocity profile and higher turbulence typically found in suburban and urban situations an effective wind speed was determined in wind tunnel studies, defined as the mean plus 1.5 times the root-mean-square of the turbulence fluctuations.

The authors estimated that occurrences of once per week, once per month and once per year corresponded approximately to probabilities of 1.5%, 0.3% and 0.02%.

C.4   Lawson and Penwarden (1975)

Lawson and Penwarden (1975) proposed the following criteria.

Activity	Beaufort No.	Probability
Acceptable for covered areas	2 or higher	<4%
Acceptable for standing areas	3 or higher	<4%
Acceptable for walking	4 or higher	<4%
Unacceptable	6 or higher	>2%

C.5   Penwarden and Wise (1975)

Penwarden and Wise (1975) published data obtained at a number of developments in England and found that the majority of locations where winds were sufficiently uncomfortable to prompt the management to take remedial action, were locations where the mean speed exceeded 5 m/s for 20% of the time or more. All locations where the winds did not exceed this criterion had no remedial action taken. This is a useful benchmark.

C.6   Gandemer (1975)

Gandemer (1975) described using a comfort parameter ψ defined as

$$\psi = \frac{\overline{U}(1+I)}{\overline{U}_r(1+I_r)}$$

where I = turbulence intensity and the subscript r denotes values at a reference point at the same height as the point of measurement of wind velocity. Either this comfort parameter was used directly or in other instances to evaluate the frequency of occurrence of discomfort was determined based on the criterion of $\overline{U}(1+I) > 6$ m/s.

C.7   Hunt, Poulton and Mumford (1976)

Based on wind tunnel tests on human subjects Hunt, Poulton and Mumford (1976) proposed the following criteria:

Activity	$\overline{U}$, m/s	$\hat{U}$, m/s	Probability($> U$)
Tolerable conditions and unaffected performance	6	9	<10%
Safe and sure walking	9	13	<1%

The gust wind speed was defined as the mean plus three standard deviations.

C.8   Melbourne (1978)

Melbourne (1978) wrote an informative review of the various criteria that had been proposed at that time including his own. His criteria are based on the following maximum wind speeds allowed not more than once per year. The gust speed was taken as mean plus 3.5 times the root-mean-square of turbulence fluctuations.

Activity	$\overline{U}$, m/s	$\hat{U}$, m/s	Probability($>U$)
Stationary, long exposure	5	10	once/yr
Stationary, short exposure	6.5	13	once/yr
Walking	8	16	once/yr
Unacceptable for any activity	11.5	23	once/yr

Melbourne adapted these criteria to other frequencies of occurrence by assuming a particular form for the probability distribution P, i.e. the Weibull distribution with a fixed value of exponent k, and forcing the probability to match the tabulated values at a probability corresponding to one event per year. Melbourne favoured using daylight hours only when counting the number of hours in the year since people tend not to use pedestrian areas as much at night. His estimates of the relationship between frequency of occurrence and probability was as follows

Number of storms per year during which $\overline{U}$ is exceeded	Probability of exceeding an hourly mean speed $\overline{U}$, $P(> \overline{U}$)	
	All hours	Daylight hours
1, once per year on average	0.025% - 0.05%	0.05% - 0.1%
12, once per month on average	0.3% - 0.6%	0.6% - 1.2%
52, once per week on average	1.5% - 3.0%	3.0% - 6.0%

C.9   Murakami, Iwasa and Morikawa (1986)

The criteria of Murakami et al (1986) were based on a two year survey of residents living near a high rise building in Tokyo combined with simultaneous wind measurements. The criteria are expressed in terms of the daily maximum wind speeds.

Criteria for wind environment based on occurrence frequency of daily maximum gust speed					
Class	Effect of strong wind	Areas applicable (example)	Level of assessment of strong wind and acceptable exceedance frequency (at height of 1.5 m)		
			Daily maximum gust speed (m/s)		
			10	15	20
			Daily maximum mean speed (m/s)		
			10/GF	15/GF	20/GF
1	Areas used for purposes most susceptible to wind effects	Shopping street in residential area, outdoor restaurant	10% (37 days per year)	0.9% (3 days per year)	0.9% (3 days per year)
2	Areas used for purposes not too susceptible to wind effects	residential area, park	22% (80 days per year)	3.6% (13 days per year)	0.6% (2 days per year)
3	Areas used for purposes least susceptible to wind effects	office street	35% (128 days per year)	7% (26 days per year)	1.5% (5 days per year)

Note: GF = gust factor at height 1.5 m, averaging time 2 to 3 seconds. In high wind speed areas typical values are in the range 1.6 - 2.5. More typical values in cities are in the range 2.0 to 3.5.

C.10   Durgin (1997)

Durgin proposed velocity criteria which vary in a continuous fashion with the probability of exceedance. The functional form of criterion wind speed as a function of probability is assumed to be of Weibull form. This then leads to the following expression for wind speed criterion as a function of probability

$$U_{equiv} = U_r (-\ln P)^{\frac{1}{k}}$$

where U_{equiv} = equivalent average wind speed, U_r = constant with dimensions of speed that depends on comfort category, k = Weibull exponent, and P = probability. Durgin assumed a representative value of 2 for k. Also the equivalent average velocity is defined as the highest of the following: the average wind speed divided by 1.103; the effective gust (mean + 1.5 × RMS) divided by 1.434; and the peak gust divided by 1.875. The criteria are then expressed in terms of the following values of U_r.

Comfort Category	U_r, m/s
1. Comfortable for long periods of standing or sitting	2
2. Comfortable for short periods of standing or sitting	2.85
3. Comfortable for walking	3.7
4. Uncomfortable for walking	4.8
5. Dangerous and unacceptable	5.3

If the actual wind speed versus probability curve differs from a Weibull distribution with $k = 2$, then the location in question may pass the criterion at one probability level and fail at another, which may be confusing to lay persons. Durgin suggested that if a single probability level is selected at which to evaluate conditions then a probability of about 5% was the optimum.

C.11   Soligo, Irwin, Williams and Schuyler (1997)

Soligo et al proposed a comprehensive set of criteria aimed at including thermal effects as well as the effects of wind force on pedestrians. In the context of this appendix only the wind force component will be described. They selected 20% probability of exceedance, i.e. 80% probability of non-exceedance, as the level at which to set the wind speed criteria for comfort. For safety it was set at 0.1%. The selection of this probability level was based on their experience that this was most easily accepted by lay persons. They used a somewhat similar concept of equivalent average speed to that of Durgin (1997). The equivalent speed is defined by Soligo et al as the higher of the mean wind speed and the peak gust speed divided by 1.85. The criteria are given in the following table.

Activity Category	Gust Equivalent Mean (GEM) speed range, km/hr	Probability
Sitting	0 - 9	≥80%
Standing	0 - 14	≥80%
Walking	0 - 18	≥80%
Uncomfortable	> 18	≥20%
Severe	≥52	≥0.1%

C.12 Durgin (2002)

Following his 1997 publication Durgin undertook further investigations into his criteria and also compared a large number of other criteria by various researchers. Based on this he has published revised criteria Durgin (2002). In these criteria he recommends using 2.6% probability as the determining level for wind speeds. The revised criteria are tabulated below.

Probability	2.6%	2.6%	2.6%	2.6%	0.01%
Category Boundary (see section C.10)	37257	37289	37318	37350	Dangerous
Equivalent Average (m/s)	3.79	5.5	7.02	8.98	16.02
Average (m/s)	4.18	6.06	7.74	9.91	17.67
Effective Gust (m/s)	5.44	7.88	10.06	12.88	22.97
Peak Gust (m/s)	7.11	10.3	13.15	16.84	30.03

Index